Sustainable Aging

Deutsche Gesellschaft für Internationale
Zusammenarbeit (GIZ) GmbH
Editor

Sustainable Aging

Cases and Cooperation in China and Germany

 Springer

Editor
Deutsche Gesellschaft für Internationale
Zusammenarbeit (GIZ) GmbH
Beijing, China

Translated by
Anne-Laure Maddy
Praia, Cabo Verde

Bridget Rooth
North Point, Hong Kong

ISBN 978-3-662-69138-0   ISBN 978-3-662-69139-7   (eBook)
https://doi.org/10.1007/978-3-662-69139-7

This Springer imprint is published by the registered company Springer-Verlag GmbH, DE, part of Springer Nature.
The registered company address is: Heidelberger Platz 3, 14197 Berlin, Germany

If disposing of this product, please recycle the paper.

Foreword

According to the World Population Prospects of the United Nations 2017, the global population is experiencing rapid and continuous ageing. This demographic phenomenon has been triggered by reduced fertility levels combined with a rise in life expectancy. China and Germany are among the countries facing a relatively intense ageing process. China has already been in this phase of demographic transition for two decades. With such a swift pace of ageing, unsurpassed by any other nation, China is heading towards becoming an aged society within less than 10 years. Germany has also been a steadily ageing society for decades. After Japan, Germany is the oldest OECD country.

The implications of population ageing are significant, if not dramatic and will affect the whole social and economic systems of a country. With an increasing dependency ratio—in other words a rapidly shrinking working population—ageing societies are losing the benefits of the demographic dividend which may lead to a deteriorated global competitiveness. Ageing is exerting a severe pressure on the whole welfare system. This especially applies to the social security system and the health sector. Similarly, a drop in tax income, induced by the falling demographic bonus, could also pose another challenge regarding providing financial resources for social protection and urgent reforms.

For over two decades, the International Community has been granting great attention to the matter of demographic change and ageing. A fundamental document and guideline that strives to offer valuable insights on the ageing society and ways to encourage the participation of older persons in the 21st century is the "*Political Declaration and Madrid International Action Plan on Ageing*", which was adopted by the Second World Assembly on Ageing of the UN 2002. The Plan focuses on the need to promote health and well-being throughout the life course as a foundation to healthy ageing. From 2015 to 2030, the World Health organisation (WHO) is running a campaign to highlight Healthy Ageing as the process of developing and maintaining the functional ability that enables well-being in older age. The Global Strategy and Action Plan on Ageing and Health (2016–2020) was adopted by the WHO with five strategic objectives including the development of age-friendly environments, the alignment health systems so that they meet the needs of older populations and the development of sustainable and equitable systems for providing long-term care. The strategy and action plans respond to almost all the 17 Sustainable Development Goals (SGDs) to be met by 2030. For example, SDG 3 sets out to "ensure healthy lives and promote well-being for all at all ages through universal health coverage including financial risk protection" to promote sustainable development and to ensure that no one is left behind.

Ageing has also emerged as a very relevant topic of the intergovernmental forum, the Group of Twenty (G20 Agenda) over the past years. During the Turkish Presidency of the G20 in 2015, principles of the "silver economy and active ageing" were formulated to address all issues of the older generation. Japan, as the so-called super-ageing society, has outlined ageing as one of the top priorities of the G20 Agenda as part of its 2019 presidency.

Germany has built a sound and consistent system of old-age provision composed of social protection, intergenerational solidarity, health care, education, and employment schemes. The country endorses an interactive, multi-stakeholder approach with the state, municipalities, communities, healthcare sector, research, the economy as well as the civil society, welfare organisations, and charitable institutions in particular. However, Germany is confronted with a significant shortage of skilled labour, a rather unsustainable social security system in the long run and higher private voluntary personal contributions to finance social protection, such as pension, health care, and long-term care schemes that may lead to old-age poverty. To cope with the demographic challenges, Germany introduced several relevant reforms to promote longer working lives such as by gradually raising the statutory age of retirement to 67 years and to facilitate phased retirement. The past years have also seen the publishing of several strategic policy documents by the German Federal Government within the aim of promoting self-determination, participation, inclusion, and dignity of the aged population. In addition, the fields of age-appropriate housing and living, care services and education as well as care insurance for securing long-term care are addressed in policy documents and schemes. Hence, this presents just how critical the German government deems these different topics as they are all embraced by the state's strategy to address the ageing challenge. To deal with ageing, Germany has successfully adopted a participatory approach with governments, municipalities, associations, trade unions, and civil organisations acting as the main stakeholders.

China has carried out a number of programmes and activities, mainly in the last eight years, to cope with the demographic shift and to improve the living conditions of the older generation: strict family planning strategies were given up, improved social welfare systems, which include long-term care insurance, were introduced, and the investment in the old age or silver industry as well as in elderly care services was diversified. The most comprehensive national strategy, the Healthy China 2030 Blueprint promotes inter alia a healthy, inclusive, and active ageing. One of the key targets of the informal and institutional elderly care is to improve the quality of the care services, such as by offering better education and training to care personnel. Compared to other countries, despite the speed of this demographic change, China did not concede a lengthy process to adjust. Thus, similar to Germany, a participatory approach has been selected to invite all stakeholders of the society to support a sustainable and equitable ageing.

Among the high-ranking dialogues between the German Federal Government and the Chinese State Council in the field of health, social security, and future of work, the questions of demographic change and ageing have been included. During the 5th Sino-German Governmental Consultations of 2018, it was agreed to cooperate in the field of "Healthy Ageing", based on a new Health Care Framework Plan. Another agreement was reached between the Federal Ministry of Education and Research and the Chinese Ministry of Science and Technology inter alia in the field of smart services, which shall include intelligent long-term care services (Care 4.0). In addition, there are numerous potential cooperation topics between China and Germany, such as old-age employment schemes, social security reforms, long-term care insurance systems, vocational education and training for elderly care personnel, and the care industry.

The Deutsche Gesellschaft für Internationale Zusammenarbeit (GIZ) GmbH acknowledges the importance of international experience exchanges in finding solutions for these social and human challenges, working on the topics of demographics, health, social protection, family, education, and youth as well as ageing. In Asia, GIZ is working in the context of several projects in the field of demographic transition with local partners, such as social security systems (health insurance, old-age insurance) in India, Indonesia, Vietnam, the Philippines, Bangladesh, and Cambodia. In 1982, the GIZ took up the work in China. Following the turn of the latest century, GIZ started to play a greater role in supporting projects related to social and healthcare issues. The social and healthcare policy dialogues were implemented in cooperation with the Social Policy Department of the National Development Reform Commission (NDRC) and the National Health and Family Planning Commission now known as the National Health Commission. These dialogues were intensified in the framework of the Public Policy Dialogue Fund with expert hearings and workshops on demographic change, inclusion, labour migration, and the ageing population in urban areas. Since then, several projects are implemented by the GIZ in the field of demographic change, including the topics of health, vocational training, medical technologies and product safety of medical equipment, and age-friendly cities and communities.

Eschborn, Germany   Christoph Beier
                                          Former CEO at Deutsche Gesellschaft
                                              für Internationale Zusammenarbeit
                                                                (GIZ) GmbH
                                                    christoph.beier@giz.de

Editors' Note

On behalf of the German government, Deutsche Gesellschaft für Internationale Zusammenarbeit (GIZ) GmbH has been supporting sustainable, social, and environmental development in China—based on global standards and norms—for 40 years. This work is in the interests of the two countries and takes account of China's changed role in the world.

To reach a broad audience, GIZ takes different approaches in this cooperation, e.g. through the publication of books with different foci. The current book *Sustainable Aging. Cases and Cooperation in China and Germany* presents policies, research, practical solutions, and development partnerships in collaboration with the private sector in the fields of elderly care education, long-term care insurance, and age-friendly cities. It is part of the book series edited by the Sino-German Centre for Sustainable Development commissioned by the Federal Ministry for Economic Cooperation and Development (BMZ).

We would like to thank all the authors for their valuable contributions most of all the BMZ for kindly providing the funding for the publications both in English and in Chinese. We are especially delighted and honoured that the German Health Minister from 2018 to 2021, Jens Spahn, wrote an introduction to this book, highlighting the importance of international exchange. A special thanks goes to the colleagues and team members who provided valuable support for this project.

We wish you an interesting read and inspiring insights!

Beijing, China

Marie Peters
Sabine Porsche
Mingming Wang
Yi Wang
Deutsche Gesellschaft für
Internationale Zusammenarbeit (GIZ)
GmbH

Personal Contribution

"I would prefer not to"—the famous, constantly repeated, defensive statement by Bartleby, the employee in Herman Melville's 1853 classic novel of the same name, also applies to our attitude in Germany towards questions of age, sickness, and becoming helpless. We would prefer not to think or talk about such issues, both in private and as a society. We avoid, as far as possible, discussing with our parents how the family would organise assistance if worst came to the worst. And we also suppress the question of who will ultimately look after us, when it is our turn. Particularly from the perspective of Germany's so-called baby-boomers, those born in the high birth-rate years of the 1960s, who are so numerous and yet have produced relatively few offspring, this suppression is reckless. Indeed, in spite of splendid medical progress, one thing is clear: the heavy blow that needing long-term care signifies for a person and his or her relatives is not something we can banish through reforms. What we can and want to do, however, is to make our assistance as comprehensive as possible. And this is why we specifically have the long-term care insurance. Nevertheless, the task is increasing. Allow me to give you just three figures: today, there are almost 4 million persons in need of long-term care in Germany; 1.7 million persons in our country already have dementia, and every year the number increases by 40,000.

There is therefore no denying the need for action. At the same time, not unlike "globalisation" and "digitalisation", "demographic trends" is, disturbingly, on its way to becoming a plastic term dangerously close to a hollow, inconsequential expression. This is dangerous simply because it is possible that we will tire of hearing the term, even before the necessary societal steps have been made and the requisite political decisions taken. That we might change course before we have even fully embarked on the process of change. Indeed, we will need change in all areas of life if we are to shape this development actively and effectively, rather than passively enduring it.

We have been told often enough how comprehensive the effects and the changes will be—also beyond politics and social systems—from the ageing-appropriate design of everyday consumer goods and the recruiting of skilled personnel, to the duration of the green-light phase at pedestrian traffic crossings and the ageing-appropriate development of city districts. Anyone can vividly imagine what it will mean for our institutions and infrastructures when they have to cater for or adapt primarily to senior citizens. However, at the same time, it would be totally incorrect to claim this was all problematic. If we use this as an incentive to make good changes, it could also be a great opportunity: living well ever longer!

Empty soapbox speeches represent one danger. The other is an approach to policy that believes that society can be held together through untenable promises of social and retirement benefits—promises that cost billions to be borne by our young people, who will be unable to pay. Above all, we may by no means reduce the level of reflection attained thus far, nor back-pedal on the political consequences already drawn, such as raising the retirement age—especially in light of a trend that will soon make the current ratio of two social security contributors to one pensioner seem like a golden age. In fact, we should be consistently linking the pension entry age to the rising rate of life expectancy. Certainly, if we continue to live longer, we will also have to work longer to be able to finance this age gain.

The Federal Republic's first great Chancellor, Konrad Adenauer, today also stands for one of those fundamental errors of judgement that we human beings commit, time and time again, based on what is considered self-evident in our respective time. "People will always have children" was his conviction with regard to the sustainability of a pension system financed on a pay-as-you-go basis. Since, sadly, this has not proven to be sufficiently accurate, precautions need to be taken to ensure that, as we move towards 2050, the financing of our pension system really is secure. We would literally have to save more—both for private insurance coverage and in the building of a demography reserve—and, in the process, provide relief to parents, depending on the number of children they raise.

Indeed, what Oswald von Nell-Breuning, one of the founders of Catholic social thought and its reflections on solidarity and justice in modern societies, said in 1980 at the age of 90, still rings true today: "Those who pay contributions do not get their own contributions back when they get old. With their contributions, they do not earn their own pension; instead, with these contributions they repay what the previous generation has provided for them. In paying their own contributions, they have paid their dues and are quits. They earn the pension that they themselves wish to draw by raising children. Anyone who makes no contribution is at a major handicap". In our pay-as-you-go system, the old receive the money from the young—even if it is only from other people's children! And I am saying this very consciously, as someone without children of my own, who is prepared to make a greater financial contribution to rendering our system future-proof.

We have the same issue of fairness in the long-term care insurance. Here too, parents also raise future contributors and, in so doing, secure the system for the future. Therefore, in Germany, the contribution rate to the statutory long-term care insurance is currently already 0.25 percentage points higher for childless persons than for insured persons with children. This basic principle is proper and exemplary. In addition, every year 1.4 billion euros is set aside in a long-term care provident fund.

As a result, long-term care is the only branch of the social insurance in Germany that has a built-in provision for the future. However, here too, we could do much more so that, even after 2030, enough money will be available when the baby-boomers enter retirement, especially since no small number of them will also develop long-term care needs in the mid-21st century.

The time is now for us to reorganise things in a manner that is equitable for all generations; otherwise the decreasing number of young people in the decade after the next will find ways to shrug off or avoid their financial overload. We all recognise that we will need more money for additional professional caregivers, who we need to be able to pay better, and for support in the home. How are we going to finance long-term care in the future within the triangle of tax revenue, insurance benefits, and individual co-payments? I would like to take the opportunity, offered by the fact that the debate on long-term care is finally gaining momentum in Germany, to have an open and honest debate on how we can continue to be a humane society and maintain our social institutions, when one in three persons in Germany is older than 60 years of age and less than one in five younger than 20. This is one of those discussions we must conduct more courageously, if we wish to take ourselves seriously as a democracy of intelligent debate, able to shape its own future.

In health policy, we are now smack in the middle of seeking solutions to the problem of providing care in structurally weak areas that are affected by migratory outflows and ageing. In many cases, we will only succeed if we make bold use of digital options. Telemedicine, online consultations, distance treatment, and e-prescriptions—these are all developments that will help us and that we are in the process of introducing. As someone who grew up in a rural area, I certainly find the recent suggestion by economists that we "close villages" a scenario that must be avoided.

At the same time, there are a number of things that have changed for the better in Germany in recent years. Some years ago, it was necessary to insist tirelessly that we still need and will increasingly need the elderly and their experience, both in society and in the working world. An "obsession with youth" does not seem to be as much of a problem anymore. In addition, a growing number of older people frequently work longer—by choice. Nowadays, they continue to contribute for a long time, above all, socially—as we saw only recently with the diverse assistance provided for refugees, by older persons, in particular.

One final idea in closing. As the Minister responsible for health and long-term care, I am the greatest supporter of an active economic policy that relies on growth and dynamism, instead of self-satisfied complacency. For, we must first earn what we distribute within our social security system . In this publication together with our

Chinese partners, I would like to repeat what I often say when speaking at home, as a show of willingness to engage in healthy competition: We want to be digital world champions so we can afford health, long-term care, and pensions in an ageing society. We will continue to need sustainable economic growth in the future as well. For our social peace, and for Sustainable Ageing!

Jens Spahn
Federal Minister of Health (2018–2021)
Federal Ministry of Health
of the Federal Republic of Germany

Contents

Editor and Contributors

About the Editor

Deutsche Gesellschaft für Internationale Zusammenarbeit (GIZ) GmbH is a global service provider in the field of international cooperation for sustainable development and international education work, with 22,199 employees. GIZ has over 50 years of experience in a wide variety of areas, including economic development and employment, energy and the environment, and peace and security. Our business volume is around 3.1 billion euros. As a public-benefit federal enterprise, GIZ supports the German government—in particular the Federal Ministry for Economic Cooperation and Development (BMZ)—and many public and private sector clients in around 120 countries in achieving their objectives in international cooperation. With this aim, GIZ works together with its partners to develop effective solutions that offer people better prospects and sustainably improve their living conditions.

Contributors

Maja Bernhardt Deutsche Gesellschaft Für Internationale Zusammenarbeit (GIZ) GmbH, Bonn, Germany

Deqing Bu School of Architecture and Art, North China University of Technology, Hebei, China

Zheng Chen Tongji University, CAUP, Yangpu, China

Shuo Cui Deutsche Gesellschaft Für Internationale Zusammenarbeit (GIZ) GmbH, Bonn, Germany

Ingrid Darmann-Finck Institute for Public Health and Nursing Research at University of Bremen, Bremen, Germany

Karl-Stefan Delank Department of Orthopedic and Trauma Surgery, Martin-Luther University (LMU) Halle-Wittenberg, Halle, Germany

Chun Ding Institute of World Economy, Fudan University, Shanghai, China

Christian Eichinger KSP Jürgen Engel Architekten, Berlin, Germany

Guanggang Feng Anhui University of Finance and Economics, Anhui, China

Steffen Fleßa University of Greifswald, Greifswald, Germany

Annika Fründt Deutsche Gesellschaft Für Internationale Zusammenarbeit (GIZ) GmbH, Bonn, Germany

Jian Fu Tianjin City Vocational College, Tianjin, China

Yiqun Guan Shanghai Qicheng Construction Planning & Design Company Ltd, Shanghai, China

Sha He School of Architecture and Art, North China University of Technology, Hebei, China

Florian Krins Deutsche Gesellschaft Für Internationale Zusammenarbeit (GIZ) GmbH, Bonn, Germany

Kevin Laudner College of Applied Science and Technology, Illinois State University, Normal, IL, USA

Andreas Lauenroth Medical Faculty, Martin Luther University (MLU) Halle-Wittenberg, Halle, Germany

Li Lin Deutsche Gesellschaft Für Internationale Zusammenarbeit (GIZ) GmbH, Bonn, Germany

Dan Liu School of Economics, Fudan University, Shanghai, China

Yiming Liu Shanghai Qicheng Construction Planning & Design Company Ltd, Shanghai, China

Nailin Lou Science, Technology and Industrialisation Development Centre of the Ministry of Housing and Urban-Rural Development, Beijing, China

Bei Lu Centre for Excellence in Population Ageing Research (CEPAR), University of New South Wales (UNSW), Sydney, Australia

Sonja Alves Luciano Deutsche Gesellschaft Für Internationale Zusammenarbeit (GIZ) GmbH, Bonn, Germany

Marie Peters Deutsche Gesellschaft für Internationale Zusammenarbeit (GIZ) GmbH, Bonn, Germany

Sabine Porsche Deutsche Gesellschaft für Internationale Zusammenarbeit (GIZ) GmbH, Bonn, Germany

Kathleen Schmidt TH Lübeck, University of Applied Sciences, Lübeck, Germany

Sebastian Schulz Transport Planner, General Manager, Buero stadtVerkehr Planungsgesellschaft mbH & Co. KG, Hilden, Deutschland

Stephan Schulze Medical Faculty, Martin Luther University (MLU) Halle-Wittenberg, Halle, Germany

Frank Schwartze TH Lübeck, University of Applied Sciences, Lübeck, Germany

René Schwesig Medical Faculty, Martin Luther University (MLU) Halle-Wittenberg, Halle, Germany

Astrid Seltrecht Faculty of Humanities, University of Magdeburg, Magdeburg, Germany

Birgit Teichmann Network Ageing Research, Heidelberg University, Heidelberg, Germany

Chengfang Wang School of Architecture & Art Design of Hebei University of Technology, Beichen District, China

Mingming Wang Deutsche Gesellschaft Für Internationale Zusammenarbeit (GIZ) GmbH, Bonn, Germany

Xiaolan Wu China Research Centre on Ageing (CRCA), Beijing, China

Shenmao Yang School of Architecture & Art Design of Hebei University of Technology, Beichen District, China

Luc Yao Oxford Institute of Population Ageing, University of Oxford, Oxford, UK

Siqi Yin School of Architecture & Art Design of Hebei University of Technology, Beichen District, China

Chenzi Yiyang Deutsche Gesellschaft Für Internationale Zusammenarbeit (GIZ) GmbH, Bonn, Germany

Bo Zhang School of Architecture and Art, North China University of Technology, Hebei, China

Ping Zhang School of Architecture & Art Design of Hebei University of Technology, Beichen District, China

Youyang Zhao Science, Technology and Industrialisation Development Centre of the Ministry of Housing and Urban-Rural Development, Beijing, China

Chapter 1
Introduction: Sustainable Ageing in Germany and China—The Role of International Cooperation

Marie Peters and Sabine Porsche

1.1 Sustainable Ageing and the Commitment of International Cooperation

When the COVID-19 pandemic broke out at the end of 2019, it once again became apparent that the elderly are particularly vulnerable during health crises, and preventive measures are paramount to ensure they age in good health and with dignity. First identified as the SARS-COV-2 virus in December 2019, the virus spread globally into a pandemic. By September 2020, 31 million cases had been reported in 188 countries, with 21 million infected people having recovered (John Hopkins University, 2020). The global fatality rate was particularly high among people aged 80 years or above until the end of April 2020, five times higher than the global average (UN, 2020).

Such a threat requires immediate action to protect lives but also longer-term political efforts on international, national, and sub-national levels to ensure what is referred to in this book as "sustainable ageing"—integrative and inclusive ageing in dignity for all. This chapter will first highlight some of the relevant international-level policy frameworks put in place to respond to the pandemic. This includes frameworks that set out guidelines to respond to the ageing demographic trend as well as the framework developed by the German government to guide international cooperation. Second, the chapter uses project cases of the Deutsche Gesellschaft für Internationale Zusammenarbeit (GIZ) GmbH in China and other Asian countries to introduce international cooperation actions to highlight the potential and importance of international cooperation in tackling the challenge. The last section of the chapter will introduce the areas of focus of this cooperation as well as contributions from academia and practitioners included in this book.

M. Peters (✉) · S. Porsche
Deutsche Gesellschaft für Internationale Zusammenarbeit (GIZ) GmbH, Bonn, Germany
e-mail: marie.peters@giz.de

© The Author(s) 2025
Deutsche Gesellschaft für Internationale Zusammenarbeit (GIZ) GmbH et al., (eds.),
Sustainable Aging, https://doi.org/10.1007/978-3-662-69139-7_1

1.2 Policy Frameworks Towards Sustainable Ageing

As a response to the pandemic and to trigger global action, the United Nations published the policy brief *The Impact of COVID-19 on Older Persons* in May 2020. It describes the broad effects the pandemic has on the elderly, revealing profound structural shortcomings in elderly care, elderly living situations and conditions as well as health security systems. The most important finding is that in some cases health care was denied for conditions unrelated to COVID-19. Cases of neglect and abuse in institutions and care facilities were reported; furthermore, a dramatic impact on wellbeing and mental health occurred, and traumas of stigma and discrimination surfaced (UN, 2020). The underlying structural shortcomings revealed in the report indicate that the elderly are not only more vulnerable in acute health crises, but generally form a group prone to long-term health-related deficits.

Given the evident need for global action, the World Health organisation (WHO) announced the start of the Decade of Healthy Ageing 2020–2030 on 7th August 2020 (WHO, 2020). This second action plan of the WHO Global Strategy on Ageing and Health aims to improve the living conditions of the elderly, their families and communities by bringing together governments, civil societies, international agencies, professionals, academia, the media, and the private sector. With this endeavour, the WHO provides a valuable framework for international action to react to acute and long-term health challenges, including the demographic challenge.

Another international long-term policy framework relevant in the context of health and ageing is the 2030 Agenda for Sustainable Development with its seventeen Sustainable Development Goals (SDGs), adopted by the United Nations Member States in 2015. The Agenda clearly outlines the task of creating an inclusive and equal world for all, regardless of age, sex, disability, economic status, etc. Among many other aspects, the SDGs aim to achieve improved elderly health and wellbeing, grant long-life education opportunities which foster social integration, and target the improvement of education programmes and facilities for nurses and other caregivers to the elderly. The SDGs do not merely seek decent work opportunities and thus economic stability for everyone, including the elderly, but also encourage the expansion of banking, insurance, and financial services. Insurance cover plays a crucial role in ensuring decent care services into old age. Last, but not least, the SDGs concede socially included and independent lives in own homes or communities to the elderly—so called "ageing in place" (Centre for Disease Control and Prevention, 2009). By addressing all spheres of life, the SDGs can contribute to making old age a productive, socially included, rewarding and thus meaningful period of life. The SDGs contribute to a more sustainable society and economy and ensure sustainable ageing.

Also in response to the pandemic, the German government has strengthened its focus on building resilient health systems for the management of outbreaks as well as efficient healthcare provision for all worldwide by making this aim a firm component of the BMZ 2030 reform strategy issued in April 2020, within the current international cooperation plan of the Federal Ministry for Economic Cooperation and Development (BMZ) (BMZ, 2020a). Following the One Health approach by the World Health organisation (WHO, 2017), in June 2020 the BMZ decided to establish a sub-department focusing on global health, pandemic prevention and One Health, and initiated a One Health Advisory Council in August 2020 to foster the promotion and maintenance of health through preventive measures (BMZ, 2020b, c; DIE, 2020). As a service provider implementing international cooperation projects on behalf of the German government, GIZ will respond to this strategy, acting as a catalyst to strengthen collaboration between public health sectors and private healthcare industries globally.

GIZ is also committed to contributing to the implementation of the SDGs, drawing on its experience as an established cooperation partner in developing, emerging, and developed countries, its technical expertise in a broad range of fields as well as its role as a consultant to the German government in the negotiation and implementation of international agreements. In line with the SDGs, GIZ follows an integrative approach towards economic, political, environmental, and social sustainability, including sustainable ageing. A worldwide network of partner countries—currently about 130—helps GIZ to closely interlink SDGs targets with existing and new programmes and projects, among them the target of an equal, integrated society and economy, currently challenged by population ageing in many countries.

1.3   International Cooperation Towards Sustainable Ageing in Germany and China

The focus on global health promoted by public and private sector cooperation is not new within Germany's international cooperation. In line with international policy frameworks, Germany's international cooperation has for many years been promoting a joint sustainable, inclusive process to face demographic challenges in Germany and China as well as other South-East Asian countries, such as Vietnam or the Philippines. In Europe, the demographic challenge of an ageing society has been evident for many decades, and has now reached former developing countries in Asia, such as China. Germany, among the countries with the lowest fertility rate worldwide, and China, as the country with the largest population on the planet, have experienced rapid changes in their demographic, social, and living structures in recent decades. These tidal shifts were brought about by changes in lifestyle patterns, migration, and an increasing overall life expectancy. In the case of China, the country's expeditious urban transformation, a rising middle class, and population control policies have had far-reaching and unforeseen social and demographic implications.

The demographic trends we see today are just the tip of the iceberg; the situation is expected to further intensify in the coming years. According to numbers from the UN Department of Economic and Social Affairs (UN, 2019), the number of people who are 60+ years old in China will increase by 264 million in the period between 2015 and 2050, reaching a staggering 479 million people by 2050 (State

Council, 2019b). In Germany, there will be an increase of 7.5 million people in the 60+ year-old category within the same period, reaching a total of 30 million by 2050. The 60+ cohort will account for a 35 per cent (China) or 38 per cent (Germany) share of total population respectively, by the middle of the twenty-first Century (UN, 2019). This increase of economically dependent elders, paired with a decrease of the number of economically active people, causes diverse, but often similar challenges to governments worldwide, making international cooperation a crucial and promising means to tackle them. Countries in Asia, such as China, with a shorter history of demographic change, can benefit from German experience, e.g. regarding the implementation of a long-term care insurance scheme or the dual vocational education and training for elderly care nurses (cf. the following chapters). Germany and other countries can in turn benefit from the digitalisation trend in China which fosters innovative approaches, for example, in the elderly care sector.

In the cases of China and Germany, an exchange on ageing challenges has been initiated on a national level by the health ministers of both countries. The third Sino-German Health Dialogue was held during the fifth Sino-German government consultations in Berlin in July 2018. Mr. Xiaowei Ma, Director of the National Health Commission (NHC), and Mr. Jens Spahn, German Federal Minister of Health at that time, agreed to focus on health cooperation in the Sino-German framework agreement for the period 2018–2020, to confront the challenge to health care systems caused by ageing populations. Spahn especially mentioned the collaboration in the field of long-term health insurance, among others, to be valuable for both countries.

Whereas Spahn's contribution to this book in the introductory section reveals that Germany is currently seeking solutions for the restructuring of the German long-term care insurance scheme to prepare for future demographic shifts, the experience exchange with other countries, among them China, which is currently piloting long-term care insurance, has the potential to offer new perspectives. In the case of China, it required some time before the government launched the first pilots for long-term care insurance, revealing the complexity of the topic. As Xiaolan Wu states in her contribution to the introductory section of this book, at the end of 2020 there were no official policies in place for long-term care. Her article provides an overview of China's policy development in recent years as the country comes to terms with an ageing society. Her suggestions for the design of a long-term care system make a valuable contribution to international exchange.

This book introduces joint projects between Germany, China, and other Asian countries within the scope of Germany's international cooperation, and provides deeper insight into three main topics: long-term care insurance, care education, and age-friendly cities and communities. These topics form the focus of Sino-German cooperation and are explored through contributions by academics and practitioners. Examples from Vietnam and the Philippines are also included to provide a full picture of the GIZ portfolio on demographic change in China and its neighbouring countries. The book aims to provide a knowledge source and to foster deeper exchange worldwide, as well as acting as a source of inspiration to find innovative solutions to tackle the challenges of ageing populations. In the following section, the chapter will first set the scene for the focus fields—long-term care insurance, elderly care education and age-friendly cities and communities—before giving an overview of the various contributions in the three fields.

1.4 Focus Fields in Germany's International Cooperation in China and Beyond: Care Education, Long-Term Care Insurance and Age-Friendly Cities and Communities

Besides the international exchange to find solutions for the demographic challenges, one opportunity for ageing countries is the emergence of a private elderly care industry in response to market demand. In Germany, the respective market appeared after the introduction of long-term care insurance in 1995. China has just taken the necessary steps to accelerate the elderly care service industry, including fully opening the elderly care market in 2020 (State Council, 2019a). The so-called silver economy attracts players from various sectors besides the elderly care service industry, such as real estate, insurance, medical equipment, and tourism, and is closely linked with digitalisation trends. The value of this industry is expected to nearly triple within 12 years from 2018 to 2030. In 2018, the industry was estimated at RMB 4.6 trillion (EUR 0.6 trillion), it is expected to reach RMB 9.8 trillion (EUR 1.2 trillion) in 2021 and will be worth RMB 13 trillion (EUR 1.65 trillion) by 2030 (People's Daily Online, 2020).

To reflect this trend and its inherent opportunity, the introductory section of this book highlights the role of the private sector in tackling such a demographic challenge. Throughout the book, projects implemented by GIZ, such as development partnerships in cooperation with the private sector or projects commissioned by the private sector, will be presented. Furthermore, in his article, Luc Yao, representing the sector-related company Merck, introduces how speculative design can spur dialogue across disciplines and lead to digital health innovations. The author not only shows the development of innovative and smart products for the elderly, but also depicts the possible role of digitalisation in the three main topics of this book.

1.4.1 Long-Term Care Insurance: Supporting an Inclusive Society

The financing of professional care, at home or in relevant facilities, poses a challenge for many families that is being taken up by governments worldwide. One solution is to establish a long-term care insurance scheme which, such as the case of Germany, is an independent part of the social security system. Since its introduction in 1995, compulsory social long-term care insurance in Germany has been reformed several times to meet the demands of those in need of long-term care as well as their caretakers.

In 2016, the Chinese central government introduced a long-term care insurance scheme and implemented it in 15 pilot regions. The insurance focused on providing basic and medical care for people with long-term disabilities, and by September 2020 the number of regions had already reached 49. Fourteen further pilot regions were then included, strengthening the long-term care for elderly with severe disabilities (China Daily, 2020). Based on the experience of the pilots, the Chinese central government plans to establish an independent insurance system and a fundamental long-term care insurance policy framework within the period of the 14th Five-Year-Plan (2020–2025). The experience from Germany is valuable to China as the country develops such a framework. Thus, this book provides a collection of contributions from the German experience and the present status in China.

Steffen Fleßa gives an overview of the compulsory social long-term care insurance system in Germany, which follows the same principles as the social health insurance. As Fleßa shows, the long-term care insurance is constantly being reformed to balance the conflict between the fair coverage of the costs and the growing expenditures. The author argues that the rising costs cannot be solved by an insurance but should rather be pre-emptively kept low by health promotion of individuals.

Chun Ding and Liu Dan provide an overview of the pilot scheme of care insurance introduced in 15 Chinese cities and summarise the experiences and current short-comings. They conclude that the process of establishing a long-term care insurance scheme revealed flaws in China's law-making apparatus, and that further adjustments of the political framework are needed to ensure the improvement of the scheme. Based on the evaluation of the results in the pilot cities, the authors make suggestions for the future development of long-term care insurance in China.

Prior to 2016 and the Chinese central government's implementation of the long-term care insurance scheme in pilot cities, the government had empowered local jurisdictions to design local long-term care insurance pilot programmes. The city of Qingdao, in Shandong province, designed and implemented the first pilot in 2012. Bei Lu and Guanggang Feng describe the evolvements, developments and operational implementation of Qingdao's long-term care insurance. They provide estimates for the future of long-term care insurance, based on administrative data as well as patients' total costs.

1.4.2   Elderly Care Education: The Basis for Healthy and Inclusive Ageing

Another essential topic addressed by governments worldwide is the provision of trained professionals to take care of the elderly. Both China and Germany are facing a shortage of caretakers for different reasons and are taking measures to bridge this gap. The sector still lacks occupational prestige, and termination rates of workers are high due to low pay and the social status associated with the profession. In 2018, Germany had a shortage of 15,000 qualified professionals to care for elderly citizens (Groll, 2018) and the number is expected to increase due to the rising number of senior citizens. To make the profession more attractive and to recruit trained staff, Germany has taken steps to reform the vocational education system to train professionals, as well as recruiting nurses from abroad.

In China, the challenge is different as the country experiences an even wider mismatch between demand and supply. In 2017, there were 42 million elderly in need of care. There were 500,000 caretakers and nurses working in elderly care facilities and organisations of which less than 20,000 had a degree in elderly care. It is expected that 14 million caretakers will be needed in the future (Wu & Zheng, 2018). China, therefore, plans to further strengthen vocational education to train elderly care nursing staff (State Council, 2020c). The German dual vocational training system, with a thorough mix of theoretical and practical training, has been used as a benchmark and has been adapted to local standards and rolled out at several Chinese medical colleges. This section of the book will introduce these cases. Other measures include the creation of nine new job titles related to senior care, one being *elderly competency evaluator*. Such a professional assesses senior daily living activities, mental condition and supports with drafting care plans (State Council, 2020b). The assessment results are also needed to determine the nursing home fees or the cost of home-based services. Another measure of the government is to lower entry requirements for nursing training so as to have enough nursing home directors or part- and full-time social workers trained by 2022 (State Council, 2020a).

Ingrid Darmann-Finck gives an overview of nurse education reform in Germany, approved in 2017 and in practice since January 2020. Whereas there were previously three vocational qualifications, there has been a shift to one General Nursing degree, which will be awarded to future cohorts. She gives reasons for this restructuring and provides arguments for other countries not to take Germany's path of three different vocational nursing qualifications.

Andreas Lauenroth et al. discuss the use of an age-simulation suit in nursing training so that the needs of ageing societies can be better understood and met by professionals in the field. Based on a survey, they conclude that with a specially designed age-simulation suit, the physical restrictions of people 25–30 years older can be felt. The authors therefore believe that professionals can be more empathetic towards patients after experiencing the restrictions the elderly face in their daily lives. The suit is also recommended for engineers or designers developing age-related products.

Based on results from a project in Panjin, China, Astrid Seltrecht in her article compares geriatric nursing trainings in Germany and China. She concludes that the direct transfer of nursing training modules from Germany to China is difficult due to different training systems in both countries. She proposes, however, to introduce more didactic training for nursing teachers in China to enable teachers to adjust training content depending on the field of practice and specific learning situations and ultimately improve learning results.

Lin Li and Annika Fründt present a project implemented by GIZ in Panjin, Liaoning Province, China, commissioned by Panjin Vocational College. This project puts the transfer of the German vocational education system of nurses into practice. The goal of the project is to implement a three-year course based on the German vocational education system to train elderly care nurses.

A second project implemented by GIZ, presented by Sabine Porsche and Mingming Wang, also focuses on the training of elderly care professionals at several Chinese nursing colleges based on the German dual vocational education system. Besides that, training for the management of officials from local government, colleges and care facilities were held throughout China based on experiences in Germany. These focused on such topics as quality management of care facilities and the integration of medical services into elderly care—both pressing topics in China's ageing sector.

The Beijing–Tianjin–Hebei region has worked to integrate the elderly care service industry into modern vocational education. Jian Fu presents this unique Chinese national strategic plan and introduce the key partners involved, both from industry and education. The Tianjin City Vocational College as one partner has been active in the promotion of a coordinated vocational education across the three regions and is thus a key player in aligning the training programmes.

To fight a shortage of professional nurses, Germany has been trying to recruit skilled nurses from abroad, including several Asian countries. Two project cases implemented by GIZ are introduced in the present book. In both cases, professionals are recruited from countries with a surplus of qualified experts who cannot be absorbed by the local market. Maja Bernhardt and Sonja Alva Luciano present the Triple Win Project in collaboration with the International Placement Service (ZAV) of the German Federal Employment Agency (BA) and the labour administrations of countries of origin. The goal is to place qualified nurses from five different countries, including the Philippines, in German clinics or nursing homes. In the article the selection and the placement process as well as the roles of the public and private partners are described.

Florian Krins presents another project case implemented by GIZ, where professional nurses are recruited from Vietnam. Commissioned by the German Federal Ministry of Economics and Energy (BMWi), the project aims to create greater awareness so that German employers consider hiring foreign skilled workers in the nursing and care sectors. Besides describing the collaboration model, the author details the benefits of this programme for the professional migrants themselves, for their country of origin as well as the destination country, in this case, Germany.

Sabine Porsche and Shuo Cui draw the attention to another GIZ implemented project addressing cancer, one of the biggest diseases among the elderly. This project case demonstrates the impact demographic change has on health and how, in cooperation with the private sector, medical professionals are trained to come to terms with this challenge. The goal is to train medical professionals to implement Precision Oncology, an innovative method for the diagnosis and treatment of cancer.

1.4.3   Age-Friendly Cities and Communities: Integrative, Inclusive and Liveable

A third dimension of the ageing challenge is accommodation. In many countries, there are simply not enough care facilities available to provide in-house care, whilst younger, healthier elderly, would prefer to lead an independent life at home while receiving care support. To provide opportunities for ageing at home while receiving care support, adjustments are needed in the personal environment of the elderly, as well as in the villages, towns, and cities they live in. These could include a (re)design of apartments to be barrier-free, improved community services, as well as the provision of social activity programmes aimed at the elderly. Thus, the topic of age-friendly living environments and inclusive cities and communities is discussed in the third section of this book.

Institutional frameworks denoting clear responsibilities and financial support structures are needed to address the challenge of an ageing urban population and to bolster the development of urban areas. As Marie Peters argues in her opening article to the section on age-friendly cities and communities, these institutional and political frameworks—if in place at all—are often fragmented and fail to address the full range of age-related issues in urban areas. They need further elaboration to tackle these challenges and to maximise the chance of developing integrative, inclusive and liveable cities and communities. She uses the examples of Germany and China, introducing the countries' evolving institutional landscapes and political approaches, which can be described as combining research and pilot-driven national level initiatives with local challenge-focused attempts at the municipal level, to find solutions.

Three authors highlight the role of construction and design standards in making cities and districts age-friendly. Nailin Lou and Youyang Zhao argue for adequate standards in the construction of elderly facilities. They identify a shortage of facilities and reluctant social acceptance for external care in China. They conclude that elderly migration, for instance, to what they call "migration care villages"—places for elderly care tourism—can be a chance to provide sufficient services and offer one solution towards sustainable ageing in China. Their suggestion to develop adequate construction standards for elderly care facilities can be applied to other relevant facilities, services, and even the cities themselves.

Youyang Zhao draws on this point in his article by describing the requirements for construction standards, arguing that they guarantee a sustainable and demand-oriented development of facilities. He outlines the drafting process of their statistical index system as a construction standard for elderly facilities that his division, the Elderly Service Facilities Association, under the auspices of the Centre of Science and Technology and Industrialisation Development of the Ministry of Housing and Urban–rural Development (MoHURD), has developed. He reports that so far, there are neither uniform evaluation systems of facilities nor coherent statistical definitions used in the planning of urban elderly facilities.

Christian Eichinger offers insight into an existing construction standard for the development of age-friendly districts, including the provision of elderly facilities and other amenities, such as public spaces, accessibility, thermal comfort or, more generally, a social and functional mix, which is applied in Germany and other countries, called the German Sustainable Building Council (DGNB) certification for districts. Based on the logic of the certification, he argues that districts should be developed inclusively by considering the needs of all residents, including the elderly. Thus, the certification is in line with the Sustainable Development Goals of inclusive and integrative cities and communities.

Sebastian Schulz and Chen Zheng argue that existing guidelines and regulations have so far failed to ensure barrier-free urban spaces providing convenient mobility for concerned groups. With their approach of using emotion mapping and bio-sensory data to measure perceptions of public spaces by users, they provide an extended focus on the peer groups themselves. This is seen as a chance to design urban space in a more human-centered and inclusive way, and to think beyond standards and regulations designed by planners towards those designed by the citizens themselves.

The following four contributions are attempts to propose concepts, models, or approaches for developing more human-centered and inclusive cities or communities. Ping Zhang, Chengfang Wang, Siqi Yin and Shenmao Yang's study contributes to the discussion on inclusive and active ageing by focusing on the design of barrier-free and convenient living environments for the elderly. The aim of their model is to create living environments that give the elderly a sense of social participation, and that provide entertainment, support, security, and life-long learning. They suggest the necessity of paying attention to the diversification of amenities, mobility and convenience, and the characteristics of a place and the elderly inhabiting it.

Based on three research projects in rural areas in Germany, Kathleen Schmidt and Frank Schwartze give recommendations for neighbourhood development concepts that strive to promote independent living and wellbeing into old age. These include considering the diversification of elderly society, the need for a functional mix, providing alternative forms of housing, considering regional differentiation, and opportunities and challenges of digital infrastructure. By applying these aspects, the authors see a chance to shift towards more inclusive and open planning policies. They argue that their results are equally important for China, where rapid economic development has wrought social change, and where the authorities could leapfrog the exclusive planning stage to focus on planning inclusive neighbourhoods. However, professional training in the central occupational fields, such as architecture, urban planning and social work, would be required to bring about this change.

Yiqun Guan and Yiming Liu in turn argue for a more integrated and people-focused approach in planning, in their case, elderly care communities. They call for a design approach of communities in China that takes into consideration the demands of a diversifying society. The authors do not see the current approach as representing changing lifestyles and consumer trends, the rural–urban divide or regional differences. Instead, they state that elderly community design should be participatory and inclusive, meeting the needs and demands of a diversified elderly society to encourage independent and spatially integrated living concepts.

In their contribution, Sha He, Deqing Bu and Bo Zhang draw our attention to elderly care facilities by comparing the evolution and development of layout designs in the United States, Germany, and Japan, drawing conclusions for the Chinese situation. They propose to implement a layout design for China that is suitable for regions with different levels of economic development; to consider open and integrated designs to avoid elderly dissociation from society; to follow certain technical design standards to increase social interaction; to improve nursing care, as well as suggesting paying more attention to the emotional needs of the elderly. Their proposals pave the way for "elderly care with Chinese characteristics".

Chenzi Yiyang focuses on the implementation level by sharing insight into an international cooperation project implemented by GIZ and the Asian Development Bank in 2016, within the framework of the Cities Initiative for Asia (CDIA). The project supported the municipality of Yichang in Central China to improve their elderly care system. With the help of CDIA, the feasibility of using private sector funding in the development of selected pilot projects in the field of elderly care was assessed. However, one conclusion is that even though private investment can be an option to lower financial pressure on the municipalities, municipal subsidies for construction and operation of services are required to guarantee high-quality services at low prices.

References

BMZ, Federal Ministry for Economic Cooperation and Development. (2020a). *Development Ministry unveils "BMZ 2030" reform strategy*. Retrieved from https://www.bmz.de/en/press/aktuelleMeldungen/2020/april/200429_pm_09_Development-Ministry-unveils-BMZ-2030-reform-strategy/index.html. Accessed 26 Dec 2020

BMZ, Federal Ministry for Economic Cooperation and Development. (2020b). *develoPPP.de. Sustainable business initiatives in developing countries*. Retrieved from https://www.develo pppde/en/. Accessed 26 Dec 2020

BMZ, Federal Ministry for Economic Cooperation and Development. (2020c). *Public-private synergies for health in times of Covid-19*. Retrieved from http://health.bmz.de/events/In_focus/pub lic_private_synergies_covid-19/index.html. Accessed 26 Dec 2020

Centers for Disease Control and Prevention. (2009). *Healthy places terminology*. Retrieved from https://www.cdc.gov/healthyplaces/terminology.htm. Accessed 26 Dec 2020

China Daily. (2020). *China expands long-term care insurance scheme trials*. Retrieved September 19, 2020, from www.chinadaily.com.cn/a/202009/19/WS5f65fb58a31024ad0ba7a969.html. Accessed 26 Dec 2020

DIE, Deutsches Institut für Entwicklungspolitik. (2020). *Anna-Katharina Hornidge appointed to the One Health Advisory Board of BMZ*. Retrieved August 28, 2020, from https://blogs.die-gdi. de/2020/08/28/anna-katharina-hornidge-appointed-to-the-one-health-advisory-board-of-bmz/. Accessed 26 Dec 2020

Groll, T. (2018). *Der Pflege gehen die Kräfte aus*. Die Zeit. Retrieved May 04, 2018, from https://www.zeit.de/wirtschaft/2018-04/fachkraeftemangel-altenpflege-deutschland-statistik. Accessed 26 Dec 2020

John Hopkins University. (2020). *Coronavirus Resource Center*. Retrieved from https://coronavirus. jhu.edu/map.html. Accessed 20 Sept 2020

People's Daily Online. (2020). *China's elderly care market boosts rapid industrial develop-ment*. Retrieved December 03, 2020, from en.people.cn/n3/2020/1203/c90000-9795499.html. Accessed 28 Dec 2020

State Council, State Council of the People's Republic of China. (2019a). *China sees improved elderly care system*. Retrieved April 01, 2019, from http://english.www.gov.cn/news/top_news/2019/04/01/content_281476589973576.htm. Accessed 15 Dec 2020

State Council, State Council of the People's Republic of China. (2019b). *China's elderly popula-tion to reach 487m around 2050*. Retrieved October 09, 2019, from http://english.www.gov.cn/statecouncil/ministries/201910/09/content_WS5d9dc831c6d0bcf8c4c14c66.html#:~:text=China's%20elderly%20population%20to%20reach%20487m%20around%202050,Office%20of%20the%20National%20Working%20Commission%20on%20Ageing. Accessed 15 Dec 2020

State Council, State Council of the People's Republic of China. (2020a). *Highlights of China's elderly care over the past five years*. Retrieved October 20, 2020, from http://english.www.gov.cn/news/topnews/202010/29/content_WS5f9a744dc6d0f7257693e9d4.html. Accessed 15 Dec 2020

State Council, State Council of the People's Republic of China. (2020b). *China offers 8 million beds for elderly care: Ministry*. Retrieved December 28, 2020, from http://english.www.gov.cn/archive/statistics/202012/28/content_WS5fe91537c6d0f72576942908.html. Accessed 26 Dec 2020

State Council, State Council of the People's Republic of China. (2020c). *China strengthens compliance supervision of elderly care services*. Retrieved December 30, 2020, from http://english.www.gov.cn/news/pressbriefings/202012/30/content_WS5febb508 c6d0f72576942aa5.html. Accessed 25 Jan 2020

UN, United Nations. (2019). *World Population Ageing 2019: Highlights*. Retrieved from https://www.un.org/en/development/desa/population/publications/pdf/ageing/WorldPopulationAgeing2019-Highlights.pdf. Accessed 26 Dec 2020

UN, United Nations. (2020). *Policy brief: The impact of COVID-19 on older persons*. Retrieved from https://www.un.org/sites/un2.un.org/files/un_policy_brief_on_covid-19_and_older_persons_1_may_2020.pdf. Accessed 29 Jan 2021

WHO, World Health Organisation. (2017). *One health*. Retrieved from https://www.who.int/news-room/q-a-detail/one-health. Accessed 26 Dec 2020

WHO, World Health Organisation. (2020). *Decade of healthy ageing 2020–2030*. Retrieved from https://www.who.int/docs/default-source/decade-of-healthy-ageing/final-decade-proposal/decade-proposal-final-apr2020-zh.pdf?sfvrsn=a1cca090_4. Accessed 26 Dec 2020

Wu, Y., & Zheng, Q. (2018). *Trends in integrated medical care and medical services in China. Exploring "the last mile" of healthy ageing*. Retrieved from deloitte-cn-lshc-the-last-mile-of-senior-care-en-181024.pdf. Accessed 26 Dec 2020

Marie Peters is an urbanisation expert with more than 10 years of work experience in East Asia, Southeast Asia and Europe. She trained as an urban geographer at the University of Cologne, Germany, with a focus on sustainable urbanisation, climate-risk resil-ience, and urban plan-ning and development. She holds a Ph.D. in Geography from the University of Cologne. Her Ph.D. research covered Chinese inter-city competition for talents in the mega-urban region of the Pearl River Delta, China. She has also published on international migration and climate adap-tation as well as on the dynamics of industrial upgrading. Her general academic interests in-clude the role of migration within urban transformation and climate adaptation. Marie joined Deutsche Gesellschaft für Internationale Zusammenarbeit (GIZ) GmbH in 2015, where she has been working as a policy advisor for different projects related to urbansiation and transport in Germany and China. She is currently based in Berlin and serves as advisor to the German Council for Sustainabile Development.

Sabine Porsche graduated in Cultural Anthropology from Philipps-University in Marburg. She joined Deutsche Gesellschaft für Internationale Zusammenarbeit (GIZ) in Beijing in 2017, where she heads the team Society and Labour. Topics of interest are el-derly care and education. From 2007 to 2016 Sabine held different positions at Tongji University in Shanghai. She held the position as German Vice-Director of the Sino-German University of Applied Sciences and besides established the DAAD Career Academy.

Part I
Demographic Change and Its Challenges in China and Germany

Chapter 2
Thoughts on China's Long-Term Care Development Policy

Xiaolan Wu

China cannot escape the ageing of its society in the twenty-first century. At the end of 2019, the country's elderly population of 60-year-olds and over-reached 254 million, accounting for 18.1%[1] of the total population. According to research forecasts in the "National Strategic Research on the Ageing Population",[2] China's elderly population is expected to reach over 300 million by 2025, accounting for 21.1% of the total population, where one in five people will be elderly. By 2050, the elderly population is predicted to peak at 483 million. By that time, one in three people will be elderly. Meanwhile, the oldest-old sub-group will continue to grow. By 2020, China's oldest-old population is expected to reach almost 30 million, and in 2050, it is predicted the sub-group will break the 100 million barrier. With a backdrop of an ageing population, chronic diseases, disabilities, and a smaller family structure, long-term care has become, on a global scale, a policy, social, and economic issue.

2.1 Policy Developments

In mainland China there is still no official policy framework in place for long-term care. At present, responsibilities fall separately on various departments working on development plans for the elderly, elderly care services and policies relating to the

---

[1] National Bureau of Statistics (2019). National Economic and Social Development Statistical Report, 28th February 2020.

[2] Ageing Population and Development Strategy Research Group (2014). Strategic Research on the National Ageing Population: Strategy Research on the Ageing Population and its Development. Hualing Publishing House.

X. Wu (✉)
China Research Centre on Ageing (CRCA), Beijing, China
e-mail: crcawu@sina.com

© The Author(s) 2025

Deutsche Gesellschaft für Internationale Zusammenarbeit (GIZ) GmbH et al., (eds.), *Sustainable Aging*, https://doi.org/10.1007/978-3-662-69139-7_2

elderly. If we look at the characteristics of policy over the years, China's elderly care service policy development can be mainly divided into three parts.

The first part saw the continuation and improvement of the civil administration's resettlement policy which mainly targets civil affairs and public pension institutions under the jurisdiction of the civil administration to ensure basic services for civil affairs. Policies during this period include, "Interim Measures for the Management of Social Welfare Institutions" (1992), "Basic Norms of Social Welfare Institution for the Elderly" (2000), and "Guidelines for the Administration of 'Five-Guarantee' Service Institutions in Rural Areas" (2006).

The second part witnessed the expansion and enrichment of social elderly care service policies, which principally target all senior groups and elderly care service industries, to continuously meet the growing demands for senior care. Policies and regulations with top-level design implications for elderly care services were introduced one after the other during the "Twelfth Five-Year Plan" period and include: the "Elderly Care Social Service System Construction Plan (2011–2015)" promulgated in 2011, "Law of the People's Republic of China on the Protection of the Rights and Interests of the Elderly" revised in 2012, "Several Opinions of the State Council on Accelerating the Development of the Pension Service Industry" 2013, and "Guiding Opinions on Advancing the Combining of Medical and Health Services and Elderly Care Services" 2015.

During the "13th Five-Year Plan" period, China's elderly care policies and practices entered a stage of rapid development. In 2013 the State Council issued "Several Opinions on Accelerating the Development of the Elderly Care Service Industry", which kicked off the first stage of an intense introduction to elderly care service policy. Starting in 2014, many ministries began intensively introducing various implementation policies promoting the standardisation of elderly care services, the cultivation of talent in the elderly care industry, and the government purchasing of elderly care services and facilities, to drive the full opening of the elderly care service market and support social capital entering the field of elderly care services. Beijing, Zhejiang, Tianjin and other areas have even adopted a method of drafting local regulations to lead and promote the development of social elderly care services and elderly home-care services. In 2016, the Ministry of Civil Affairs and the Ministry of Finance published the "Notice on the Central Government's Support for the Pilot Project of Home and Community Elderly Care Service Reform" planning that, within 5 years, the central government would spend RMB 1 billion (EUR 140 million) to support pilot projects for community home-care services. Five pilot projects have been carried out across the country. The "Thirteenth Five-Year Plan for the Construction and Development of the Elderly Care System in China" released in 2017 outlined the upgrading of home-based, community-based, institution-based, and integrated elderly care service systems as one of the country's development goals. In 2019, the General Office of the State Council issued "Guiding Opinions on Promoting the Development of Elderly Care Services", which for the first time clarified policy goals and an action roadmap for establishing a sound long-term care service system at national level.

Firstly, to establish an assessment mechanism. Improving the national unified ability assessment standards for the elderly, to carry out a unified comprehensive geriatric assessment, and use the assessment results as a basis for the elderly to receive basic care services. Secondly, to standardise service quality. Researching and establishing industry norms related to long-term care projects, standards, and quality assessment. Thirdly, to increase service supply. Upgrading the professional long-term care service system that connects homes, communities, and institutions. Fourthly, to open financing channels. Comprehensively establishing a subsidy scheme for the elderly and disabled elderly in financial distress, strengthening the link with existing subsidy policies for the disabled, accelerating the implementation of the pilot long-term care insurance system, and encouraging the development of commercial long-term care insurance products.

Entering the new era, the "Proposals of the Central Committee of the Communist Party of China on Formulating the Fourteenth Five-Year Plan for National Economic and Social Development and Long-Term Goals for 2035" adopted by the Fifth Plenary Session of the Nineteenth Central Committee of the Communist Party of China, put forward a "national strategy to respond actively to an ageing population", to "upgrade the basic elderly care service system, develop inclusive care services and mutual assistance for the elderly", "build an elderly care service system that is coordinated with home-based and community-based institutions, and integrates medical care and healthcare", as well as to "improve the comprehensive supervision system of elderly services", to clearly define the requirements for the "four beams and eight pillars" to build a functioning elderly care system.

The third part witnessed a long-term care insurance policy entering the public eye. The "Outline of the 13th Five-Year Plan for the National Economic and Social Development of the People's Republic of China" clearly requests to explore the establishment of a long-term care insurance system and carry out pilot projects for its implementation. In July 2016, the Department of Human and Social Affairs released its "Guidance on the Pilot Project of a Long-term Care Insurance System". The mission of the pilot work is to explore ways to establish and raise funds for social assistance and mutual aid, provide a social insurance system with financial or service guarantees for long-term disabled people for basic life care and basic life related medical care. Since 2016, the state has organised for some pilot cities to actively implement a long-term care insurance system and has conducted fruitful analysis in terms of institutional framework, policy standards, operating mechanisms, and management methods. In 2020, the National Medical Insurance Administration and the Ministry of Finance issued "Guiding Opinions on Expanding the Pilot Programme of the Long-Term Care Insurance System". It plans to test the results of the pilot on a larger scale and further explore the feasibility of the long-term care insurance system framework in adapting to China's national conditions.

2.2   Long-Term Care Developments

The support for national elderly care policy has been unprecedented in recent years. The scale of social investment has continued to increase, the number of service institutions and the number of beds has risen significantly, and China's elderly service industry has entered a stage of rapid development. The construction of elderly care service facilities in the country has developed rapidly, the number of elderly care beds has increased significantly, and the coverage of community elderly care facilities has continued to expand. As of the end of 2019, there were a total of 204,000 elderly care institutions nationwide, offering a total of 7.75[3] million beds, with 4795 fully equipped institutions registered as providing both medical and elderly care services. Among these, there were 3172 elderly care institutions hosting medical and health-care centres, and 1623 centres having launched elderly care services. But despite there being an additional 56,400[4] cases of medical and healthcare centres in discussions to team up with elderly care institutions, the establishment of an integrated long-term care service system is still in the initial stage of development.

2.2.1   Long-Term Care Services Remain Weak Overall

Though the number of elderly care beds is rapidly increasing, occupancy rates are relatively low. Statistics at the end of 2014 show that the national rate of empty elderly care beds was as high as 48%. The rate of empty elderly care beds in Beijing was 40–50%. Even in Shanghai, which has the highest levels of ageing, the overall occupancy rate of elderly care institutions was less than 70% (Xie, 2015). The large number of empty beds for the elderly is mainly due to the following reasons: (1) The sheer scale of newly built elderly care institutions and their location often on the outskirts of towns, causing the common phenomenon of "no beds in urban centres, and many empty ones in the suburbs". Elderly people cannot find a place in their nursing home of choice, and nursing homes that do have vacancies are too far away from home. (2) There are few long-term care institutions and too few beds. Many elderly care institutions are not geared up to receiving patients who are most in need of such care, such as the disabled elderly and those suffering from dementia, in terms of the planning and layout of the facility, architectural design and service functions, and it is difficult to effectively meet the rigid demands of the market. (3) There is insufficient effective demand. Restricted by income levels, it is difficult for the potential needs of the elderly to be transformed into actual purchasing power. The push for prime real estate and high-end elderly care projects has meant that supply

---

[3] Li, J. (2020). Implementing a national strategy to actively respond to an ageing population. Guangming Daily 6th ed.

[4] National Health Commission of the People's Republic of China (2020). 2019 Statistical Report on the Development of Medical and Health Services in China.

is not aligned with mainstream market demands, causing resource misallocation and waste.

The rate of use of community-level facilities remains low. In recent years, the construction of home-based care centres and day care centres has increased rapidly in various regions across China. Low investment, quick results, and wide-reaching benefits, mean they have become an important starting point for local governments to expand the provision of elderly care beds and meet the needs of the elderly close to home. However, the efficient and comprehensive use of such facilities needs to be urgently improved, many have become unsustainable flagship projects. The main reasons for this phenomenon are: the lack of full-time management personnel and professional service personnel, and the extremely inadequate level of professional care, rendering the service projects futile, and making it difficult to target the disabled and those suffering from dementia who really need access to such services; difficulties in recruiting new blood, and poor operating mechanisms, making the daily operations increasingly problematic; the unreasonable planning of the facilities, layout and non-standard architectural design, causing the facility to be able to cater only to healthy and active elderly people.

Finally, there is the obvious problem of fragmentation—the three major components of home, community, and institution are disconnected from each other. For a long time, the functions of home care, community care, and institutional care in various regions have been independent and developed separately, which has led to home care and community care lacking in social and professional expertise, as well as having insufficient professional social service resources. It is difficult to change the service angle from targeting empty nest elderly people in need and expanding this to cover the disabled elderly and those suffering from dementia. Meanwhile, institutional elderly care has been over-institutionalised and stand-alone, making it difficult to effectively support the development of community home-based elderly care services; it does not meet the psychological needs of the elderly in that they are separated from their familiar living environment.

In short, the number of elderly care beds can be increased in the short term, expanding care coverage, due to the sheer scale of elderly care institutions despite their location in the suburbs, as well as the opening-up of services extending access, despite the simplification of service functions, and the limited scope of day care centres and other community facilities.

But this does not translate into a simultaneous increase in the efficiency of use of beds and facilities. More importantly, it falls short of providing basic elderly care for the disabled elderly and those suffering from dementia.

2.2.2 Lagging Social Development

Firstly, the fragmentation of social services is a key issue. The elderly have a wide-ranging continuum of needs when it comes to social services, but the fragmentation of these services in China is striking. The country not only has services for different groups of people, but also different types of services each belonging to different departments. Resources are scattered, multiple systems have been established, comprehensive service capabilities are weak, and the efficient use of facilities is low, making it difficult to meet overall human-centered care needs.

Secondly, the development of communities is lagging. Community care for the elderly is dependent on the building and design of community infrastructure. It is extremely important to strengthen the development of communities, to enhance the residents' sense of belonging and identity within the community, to truly form a healthy living environment. It is only in this way that there can be an effective combining of resources inside and outside the community, whereby residents are motivated to volunteer to care for those in real need and shed light on them. Recently, there have been incidents of residents strongly opposed to the construction of community nursing homes in some regions in China. This reflects the difficulty in creating smaller models of nursing care facilities and integrating them into local communities, further emphasising the need to promote community care concurrently with the design and construction of such communities.

2.3 Policy Recommendations to Promote the Development of Long-Term Care

2.3.1 Clarifying the Attributes of Long-Term Care Within Social Services

Long-term care is a key area of research and an important social topic which has led many researchers to define the term in a similar way but from different perspectives (Chen, 2003; Kane & Kane, 1987). Long-term care is seen as a kind of social service necessity born from the inability of some people to carry out basic self-care due to partial or total loss of physical or mental functions. Appropriate services are provided according to the needs of the disabled person or their caregivers, such as physical care, assisted living, social participation and related medical services. In practice, all countries generally include long-term care within community social services. The German long-term care law stipulates that "the need for care" refers to when daily life needs to be taken care of continuously and regularly for at least 6 months due to illness or impediment. Japan's "Long-term Care Insurance Law" stipulates that the state of nursing care refers to a state that is expected to require constant care, support, and protection for part or all basic daily activities such as bathing, excretion, eating, etc. due to physical or mental disorders.

Traditionally, long-term care was delivered by family caregivers and considered a family duty. However, due to increasingly ageing populations, common changes in the spectrum of diseases and changing functions in family care, long-term care increasingly exceeds the burden of family caregivers. In this context, families in distress may fall into isolation and helplessness, at the cost of lowering quality of life, or trying their best to use medical services to make up for the lack of care, leading to community hospitalisations such as repeated or long-term hospitalisations for the disabled. Therefore, the development of long-term care aims to transform the care of the disabled elderly from a purely family-supported service into a social service, combining various efforts to provide proper care for the elderly, distinguishing it from medical services, to avoid the misuse and abuse of medical resources, whilst enabling the disabled elderly and those suffering from dementia to stay connected to normal life and enjoy indispensable quality of life.

The long-term care system is an extension of acute and subacute care in the medical system. However, planning long-term care with acute or subacute care will lead to over-medicalisation and improper use of resources. Over-medicalisation can also diminish the support role the family plays in some long-term care cases. Therefore, the long-term care system and the medical service system should be seamlessly connected to provide convenient and integrated services for the elderly, but there is undoubtedly a clear distinction between the two. Long-term care is for the disabled elderly and those suffering from chronic dementia. It focuses on basic life care and aims to maintain the functions of the disabled and improve quality of life. Medical services, on the other hand, are aimed at acutely ill patients who need treatment in the short term, focusing on medical care, aiming to cure the disease and prolong life.

2.3.2 Elderly Care Services Based on a Long-Term Strategy

From a public service and policy responsibility point of view, and in the context of an ageing society, the government should be primarily responsible for caring and actively responding to the increasingly long-term social care risk, by meeting the long-term care needs of the many disabled elderly and those suffering from dementia, and long-term care services should be regarded as the essence and main path of development to combine medical care and healthcare. Some researchers[5] uphold that the combination of medical and health care services can only exist with a long-term care perspective framework, which would ensure that there would no longer be infinite generalisations or unclear semantics in issues such as qualitative, standards and types of facilities, financial mechanisms, and teams.

Therefore, we should firstly integrate long-term care into the general discussion of social policy to lay the foundations for extensive social understanding, while gradually establishing an independent long-term care policy system with defined long-term development goals and concepts, service groups and content, along with institutional arrangements and implementation paths. The provision of long-term care and the resources allocated for training is still rather scattered, which inevitably results in a range of problems such as unclear concepts, unbalanced responsibilities,

---

[5] Dong, H. (2017). Integrated Health and Social Care Problems and Reflections in Different contexts. Wechat Official Account: Elderly Care Service Industry Research Centre.

a mismatch of resources, regional imbalances, and a lack of coordination between related systems.

2.3.3   Implementation of a Comprehensive Action Plan

Concerning pension security, we are, in our opinion, entering new territory. If you navigate in uncharted waters without first carrying out research and planning, it becomes difficult to achieve the desired results. The array of concepts currently on the table is bewildering, they are unclear in practice, there is a lack of consensus and putting it all down on paper to visualise leads to confusion.

Long-term care is a kind of personal social service with long-term, comprehensive and close-knit characteristics. To maintain a normal life for a disabled elderly person or someone with dementia is not an easy task for the family and nor is it easy for society. The sustainable development of long-term care needs the joint efforts of the family, community, society and the government, to avoid unbalanced responsibility; the three basic resources—facilities, human resources and financial resources are needed together, so as to avoid a system collapse. Demand needs to be evaluated, a referral service and tracking supervision set up to form a closed circuit to avoid mismatches in resource distribution. The health care system needs to be effectively linked to the social service system, to avoid a gap in services. It is thus also necessary to carry out overall policy planning based on thematic research, and then implement a long-term care special action plan to fully release the existing policy effects and avoid gaps, mismatches or difficulties in implementing policies and projects.

China has already started to carry out long-term care insurance pilot projects, though at this stage, the establishment of long-term care social insurance is far from being a final product of our long-term care system. In the initial development stage, diversified trials should be encouraged in all regions, where insurance payments and taxation can be collected as a supplementary source of financial support for disabled groups and those suffering from dementia, to provide appropriate benefits across the board, where funds are controlled, users have options, and resources for service and awareness in society grow simultaneously.

2.3.4   Reasonable Distribution of Infrastructure

First, a complete adjustment of the construction concept is necessary. This means moving away from a focus on elderly care bed space to focusing on long-term care bed space, away from a focus on the number of beds to a focus on bed occupancy rate, away from a focus on the elderly care facilities coverage to focusing on their usage rate. Emphasis is needed on satisfying the basic care service needs of the elderly, to improve the overall care plan for elderly services, and to establish a connected and integrated long-term care infrastructure system and service.

Second, the community needs to be embedded into the long-term care infrastructure system. Ageing in place does not mean de-institutionalisation. Since the beginning of the twenty-first century, the emphasis is no longer on whether to institutionalise or not, rather, emphasis has now shifted to meeting care needs and fulfilling them in the most appropriate way, arranging care based mainly on cost-effectiveness and health-effectiveness, to allow the industrial system to develop naturally. Moreover, the ratio of institutional to non-institutional care can be about 2:8 or 3:7 or 25:75 or even one to one, the boundaries have become blurred, and this has become a trend and a result of development worldwide (Li, 2010). Considering the cost-effectiveness of institutional care and the wishes of the elderly, community care has already become the mainstream of social welfare policies in all countries around the world. However, in the long-term care system, institutional care is still indispensable because it can provide 24-h professional care for the elderly with severe disabilities, and can also provide a professional support platform for those in home-care or community care. Therefore, it is necessary to use institutional resources in the community as much as possible, to promote the scaling down of institutions and a deeper connection with the community, as a development path for the future.

To begin with, gaps need to be filled, first starting small and then branching out. It is like planning the construction of a primary school or high school. To set up small and medium-sized facilities for long-term care in residential areas, embedded in the community, one needs to emphasise small and medium-sized institutions rather than micro-institutions to maintain the necessary scale effect and solve operating balance problems, to ensure the sustainability and replicability of the operation model of the institution. An institution providing specialised long-term care in the region will play the role of a service hub and demonstration base, making it possible for long-term care for the disabled elderly and those suffering from dementia to be provided at home, as well as driving the transformation and upgrading of existing community senior care facilities, breaking down the barriers between home, community and institution.

Next, links must be formed, and multi-position management encouraged in the spirit of mutual complementarity. Small and micro-facilities such as home-based care centres and community day care centres are mainly for the elderly, disabled and those suffering from dementia, assisting families in providing general life care, cultural entertainment, health care and other services. Small and medium-sized long-term care institutions that are embedded into communities can shift to providing care for moderate and severe conditions, especially those with severe disability and dementia, providing them with more intensive and professional long-term care. Allowing the elderly to leave their homes but not the street they live on facilitates family visits and communication, and gives them a living care space instead of just a shelter or a clinical medical care space. The Taiwan Long-Term Care Plan 2.0 highlights the cohesion and interaction between the three-level community-based long-term care institutions. It is understood that the lack of long-term care facilities and functions in the community is the root cause of the difficulty in achieving substantial breakthroughs in long-term care services and that home and community long-term care capabilities are an insurmountable part of the long-term care system.

Finally, there needs to be infrastructure protection mechanisms put in place, by firstly reinforcing land and housing security. The government should incorporate basic elderly service facilities into the overall urban planning and consider a detailed regulatory plan. The civil affairs administration should unite various departments including those responsible for land resources, planning and housing, to carry out scientific planning and rational layout to spur the construction of small and medium-sized long-term care institutions embedded in the community through new constructions, reconstruction and transformations based on the composition of the local elderly population, basic elderly care service needs, existing facilities and space resources.

Secondly, the effective operation of the facility should be ensured. The government should strictly adhere to standards and withdrawal mechanisms, and a sound regulatory mechanism. The basic elderly care facilities, including the small and medium-sized long-term care institutions embedded in the community, will be handed over to the standardised, specialised and connected private elderly service institutions for market operation without compensation or low compensation, to coordinate industrial resources, systematically foster human resources, leverage scale effects, and improve service quality.

Thirdly, establishment of a financial security system. Providing necessary financial support for long-term care for the disabled and elderly suffering from dementia can come through various channels such as charity assistance, social welfare and long-term care insurance. Thus, guaranteeing the continuous operation of long-term care facilities, and enhancing the affordability of long-term care services.

2.3.5   Creating a Comprehensive Care System

To meet all of mankind's care needs, it is fundamental to have at least three key systems in place—acute care, long-term care and community life care. Some developed countries and regions are also promoting the establishment of a system of provisions between acute care and chronic long-term care which is known as "sub-acute, post-acute or intermediate care". Only by integrating and coordinating these systems is it possible to provide complete services required by citizens, so that the elderly can still enjoy community integration care even after suffering from severe disability, and they can continue to live in their community. Since each system has its own professional barriers and operational logic, all countries around the world are striving to achieve full integration of different services for the life-long care of their citizens to achieve ageing in place.

To achieve this, one must start from improving functions within the "medical" and "social care" elements, as well as strengthening connections between the various services. Firstly, by adjusting and optimising existing medical resources, strengthening basic public health services and primary health care services, and effectively implementing detailed family doctor contract services to provide personalised medical services, health management, appointment referrals, medication guidance and other services for the elderly, while facilitating daily doctor visits and making it easier to obtain health services, thus making health services more effective and convenient. Secondly, by providing the best professional and most comprehensive long-term care services within the elderly care system as well as comprehensive maintenance and rehabilitation services for disabled elderly and those suffering from dementia. Based on life care and social participation, the service capacity of non-healing health management and rehabilitation care should be strengthened. Through the combination of medical and health care services, an effective resource integration mechanism is established to caregivers for life assistance, support, care and rehabilitation care. The third step is to improve the quality of existing elderly care services, allowing elderly care institutions to complement these, and to improve the quality of community-based elderly care services. At the same time, there should also be targeted provision of assisted living services, such as food delivery and companionship services, to ensure the safety and stability in the daily lives of the growing number of elderly people who live alone or are living with dementia. The fourth step is to strengthen institutional design connected to the system, so that the elderly can easily switch to different care systems, like, for instance, the proposal that Taiwan has made of a post-discharge preparation service to encourage hospitals to prepare patients before discharging them and to conduct follow-up after the discharge process. It is designed to assist in the referral of patients to community medical institutions, homecare, long-term care institutions and other follow-up care resources so that patients can successfully return to community life, reducing the risk of emergencies and re-hospitalisation in the short term after being discharged.

2.3.6   Establishing Multiple Partnerships

The establishment of close partnerships and sharing of responsibility between different organisations such as government, market, social and family, is needed to form social concepts and operational models for self-help, mutual assistance, communal assistance and public assistance. As a first step, government responsibilities need to be clarified. The most important element is to clarify the "basic guarantee and bottom line" in policy documents related to elderly services, this means clarifying what kind of social safety net should be built into elderly services. Faced with an ageing society, it is undoubtedly necessary to focus on elderly people, in particular those groups facing long-term care risks, to eliminate the spread of emotional fear about ageing in society and improve the effectiveness and accuracy of government funding. Based on this understanding, one needs to start from zero fully grasping the real needs, implementing overall planning; carrying out evaluations of physical conditions and assessments of economic status, implementing insurance according to different category groups; establishing basic elderly service facilities to ensure effectiveness in service delivery; integrating existing resources and opening funding channels; strengthening social dissemination and building a support environment. By doing a good job of ensuring basic elderly services, the results of promoting related industrial upgrading and market-oriented transformation will also turn out better.

Secondly, coordination among departments needs to be encouraged. To solve the lack of consensus of elderly care services, and problems from multi-channel management, unclear responsibilities, and the fragmentation of administrative institutions, stronger coordination is required among the different departments to achieve cooperation and interaction among multiple entities. Starting out with the logic of "how to solve problems better", and promoting the coordination of specific issues, different departments must focus on the long-term care service needs of the disabled elderly and those suffering from dementia within an ageing society and find their own "jurisdiction list" and "responsibilities list", gradually transforming the entire administration.

Thirdly, the relationship between the government and the market needs to be well managed. (1) Handling public welfare and efficiency issues. The functional positioning and orientation of public elderly care institutions needs to be clarified to ensure the "limitations" of their service scope and the "effectiveness" of service functions. The government also has certain responsibilities for providing inclusive elderly care services, but it does not directly produce services, instead it provides institutional provisions, financial support, supervision and management, etc., and forms a multi-strength participation for the development of the elderly care industry for mass consumption. (2) Strengthen supervision. The diversification of the main body of welfare does not mean the removal of state responsibility, but rather the strengthening of management roles. How to protect the participation of social forces, whilst avoiding accidents from happening, and protecting the rights and interests of the elderly, poses a real challenge to the government's regulatory capacity. (3) Develop the market. Based on government responsibility, measures should be adopted using appropriate guidance, encouragement, supervision, reward and penalty, to increase effective market demand, thus expanding effective supply and fostering a market environment that has an orderly development and fair competition. Government should pay attention to the development and construction of industry organisations and play an active role in providing decisions and consultation, service industry development, market regulation, and expansion of foreign exchanges. It should be emphasised that government departments, research institutions, social organisations and enterprises should be strongly involved in the discussion, formulation and evaluation of policies and promote the creation of multi-party partnerships.

Fourthly, long-term care should be integrated into the family structure. The importance of family care is indisputable, but one must also fully recognise the practical difficulties and situations faced by the family when taking care of the elderly. A community service rich in resources and capabilities, and active in the development process of home-based and community-based services can provide different types of assistance and support for family caregivers, improving the quality of life of family caregivers, strengthening the family's skills in providing elderly care and maintaining a stable and harmonious family.

Fifthly, long-term care should be integrated into community governance. Community care needs to effectively integrate resources inside and outside the community, to stimulate the capabilities and enthusiasm of community residents, voluntary organ-isations and other social forces to develop good neighbourly mutual assistance and elderly services. This must be promoted in parallel with facilities in the community. Innovation in community governance should be encouraged through "three social linkages"—the community, social organisations and social work—promoting a people-oriented community spirit, creating a community life of mutual respect and mutual care, and effectively finding and serving those in need. More impor-tantly, it is necessary to gradually establish a social care culture that is compatible with the ageing society and society as a whole, to solve the problem of long-term human-centred care, social resource integration and social inclusion.

2.3.7 Laying a Solid Foundation for a Policy Legal System

The promotion of any major work must first have correctly defined values before it can affect the relevant people and bodies to function as an effective operation model. The sustainability of such a model depends on a good social framework, and the construction of a legal system framework, which is particularly urgently needed. In addition to a legal system framework, there is still a need for a social and human quality framework that is legitimate, just, altruistic, and offers support, that is enough to implement the legal system framework to respond to the arrival of the long-term care era (Li, 2010). To promote the overall development of long-term care and face the numerous theoretical issues and problems in practice, we should first introduce the concept, essence and positioning of long-term care to society, to form a basic consensus, to then jointly encourage long-term care to take root in all of society.

An improved policy and regulation system needs to be established to regulate long-term care services and systems, to manage institutions and human resources and protect the rights and interests of service providers. At present, the long-term care policy system in mainland China is still rather lacking, and related laws are even more so. A useful reference is Japan, where the country has formed a most complete and detailed legal system, which includes the Care Insurance Law System (including the introductory law, implementing order and detailed rules for the implementation of the Care Insurance Law) as well as other relevant decree laws (such as the Health Insurance Law, Medical Law, Life Protection Law, National Health Insurance Law, Elderly Welfare Law, and Elderly Health Care Act, etc.) to provide institutional guarantees for the long-term care of the entire society.

References

Chen, C. Y. (2003). Long-term care for the elderly. *Formosan Journal of Medicine, 7*(3), 405. https://doi.org/10.6320/FJM.2003.7(3).13

Kane, R. A., & Kane, R. L. (1987). *Long-term care: Principals, programmes, and policies*. Springer.

Li, S. D. (2010). The development and promotion of "long-term care." *Taiwan Medical Community Journal, 53*(1), 49.

Xie, Q. (2015). *Elderly care facilities, beds hard to come by or low bed occupancy rate.* Guangming Daily. Retrieved June 16, 2015, from https://epaper.gmw.cn/gmrb/html/2015-06/15/nw.D11000 0gmrb_20150615_1-11.htm

Xiaolan Wu Ph.D. is the Deputy Director of the Institute of Ageing Health and Livable Environment of the China Research Centre on Ageing, member of the National Expert Committee on Barrier-Free Environment Construction, and expert of the Beijing Municipal Guidance Committee for Ageing Care Industry Construction. She has organised or participated in many projects of the national ministries and commis-sions and the National Social Science Fund of China; participated in the writing of "Actively Responding to Population Ageing with Building a Well-off Society", "Survey on the Status of Long-term Ageing Care", "China Ageing Cause Devel-opment Report", "China Ageing Industry Development Report", "Analysis of the Status of Beijing Home Care Facilities" and other books; participated in the trans-lation of the "Ageing Theory Handbook"; participated in the organisation and writing of the first domestic "China Age-friendly Environment Development Re-port" and "China Age-friendly Environment Construction Knowledge".

Chapter 3
Digital Health for an Ageing Society

Luc Yao

3.1 Challenges and Opportunities Presented by Demographic Change

Population ageing is a critical global trend that will have profound and long-lasting implications for governments, industries and society as a whole. At the same time, the world is experiencing an explosion of innovation driven by technology, digitalisation and globalisation. The combined force of these factors is unprecedented; and the challenges, opportunities and changes they will bring will be highly significant for the planet.

Population ageing results from a significant rise in life expectancy combined with a sharp decline in fertility rates. The World Health Organisation (WHO) reports that the world is accelerating towards ageing societies, as shown in Figs. 3.1 and 3.2. However, increased life expectancy does not automatically signify improvements in the areas of health, happiness and social harmony. As a result, policymakers, think tanks, NGOs and businesses are exerting much time and energy to study, invest in this topic as well as experiment with potential solutions, such as digital health tools and ecosystems, that can support healthy ageing.

3.1.1 Germany

Germany currently has almost the oldest population worldwide, second only to Japan. In the past, many demographic forecasts suggested that Germany's population would decline by more than 10 million by 2050 due to the country's low birth rate. However, the latest projections are much more positive, and the population level should remain

L. Yao (✉)
Oxford Institute of Population Ageing, University of Oxford, Oxford, UK
e-mail: luc.yao@ageing.ox.ac.uk

© The Author(s) 2025
Deutsche Gesellschaft für Internationale Zusammenarbeit (GIZ) GmbH et al., (eds.),
Sustainable Aging, https://doi.org/10.1007/978-3-662-69139-7_3

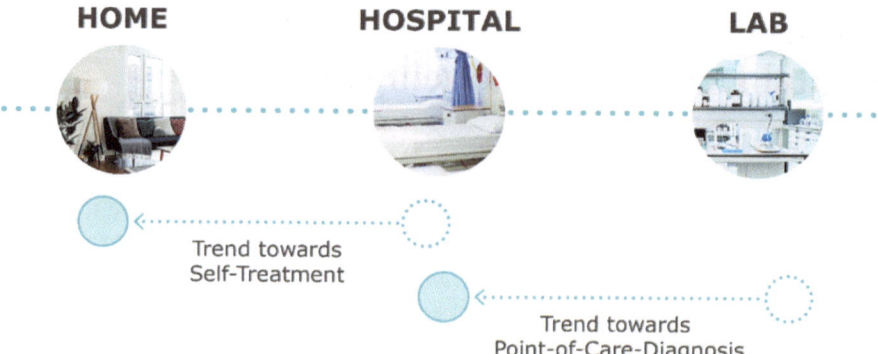

Fig. 3.1 The shift towards faster diagnosis and more self-treatment in healthcare (*Source* ma ma Interactive System Design 2017)

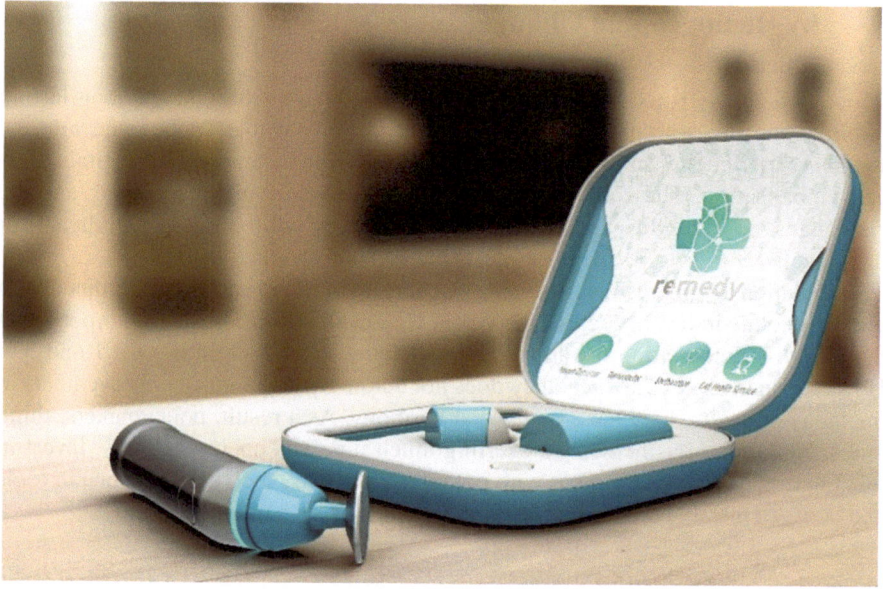

Fig. 3.2 Portable telehealth kit (*Source* ma ma Interactive System Design 2017)

above 80 million until 2060. The influx of immigrants into Germany is keeping the population stable and introducing a younger workforce into industry. With a robust economy, Germany is focusing on enhancing the wellbeing of the ageing section of its population and extending contribution that this group makes to society at large.

Population ageing brings not only challenges but also carries significant opportunities for Germany. Germany is renowned for its world-leading scientific and technology clusters, which include the pharmaceutical, nutrition, Industry 4.0 and medical device sectors. Digital health is becoming a hot topic, demonstrating substantial potential to connect expertise and assets and to keep the ageing population closely connected with the rest of society. In the future, therefore, true innovations will call for more than merely scientific and technical progress. Empathy with users, the introduction of new business models and the development of corresponding new regulations will also have a significant part to play. As a result, several industries, like finance, design and education, are, therefore, actively involved in the development of new product and service innovations that deal with the country's ageing population.

Many of the technologies, services, policies and ecosystems, which have been developed in response to the population ageing phenomenon, have the potential to be highly valuable to other countries too. Organisations such as German Healthcare Partnership (http://ghp-initiative.de/) and GIZ (Deutsche Gesellschaft für Internationale Zusammenarbeit (GIZ) GmbH) are pioneering collaborations of this kind.

3.1.2 China

With almost 1.4 billion inhabitants, China is the world's most populous nation. The average age of Chinese citizens is relatively young, but the country's population is ageing at a faster rate than anywhere in the world. This situation has been fuelled by a rapid rise in life expectancy twinned with the effect of the country's one-child policy which was enforced for 36 years until its repeal in 2016. By 2050, the dependency ratio of the population—which compares those not in the labour force with those who are working, or who can work full-time—will grow to 44 per cent. This situation poses enormous challenges for the country (Rapoza, 2017).

Firstly, China will need to take care of its elderly, and it's very likely many of the older generation will not receive the support from their offspring that was traditionally the norm according to Chinese culture. Many of the elderly tend to live in rural areas while many young people migrate to cities, and, therefore, specifically designed communities in the countryside may be able to offer these older people support. Furthermore, another obstacle arises from the fact that China will require its ageing population to contribute to the country's economic activity—otherwise, high debt and dependency ratios may become the greatest challenge to the country's economic sustainability. On top of this, China also faces growing demand to increase the wellbeing of its general population. The concentration of wealth in the hands of a small percentage of the population has already generated considerable social pressure and might even put the stability of Chinese society at risk if the issue is not tackled carefully and in good time.

On the other hand, China is rapidly developing the world's leading digital sector. Many new applications, such as electric vehicles as well as a range of e-commerce and shared economy platforms—are currently being piloted in China. Above all, the Chinese government has recognised the threat posed by the country's ageing population and provided significant policy incentives to encourage appropriate technical, business model and social developments in response to this danger. 'Healthy China 2030' is promoting healthy lifestyles, fitness, liveable cities and communities. The country is also placing greater emphasis on prevention rather than the treatment of illness. Attempts to apply digital tools that will facilitate the extension of healthcare to communities and homes are ongoing. Digital health is seen as one of the key disruptive innovations of our age. This is demonstrated by the major government initiative of 'Made in China 2025' which aims to bolster the development of relevant medical devices and digital tools. Key industries, regional authorities and academics have actively responded to these initiatives, triggering rapid advances in research, experimentation and investment.

While China is addressing the challenges and possible opportunities of its ageing domestic population, the country is also actively promoting international collaborations (Tan et al., 2018). China will need a considerable amount of cutting-edge (technical, social, know-how) expertise to design and trial appropriate ecosystems. While China gains experience for its own use, it can also benefit the world by sharing its successful technology-driven innovations across the globe.

3.1.3 A Unique Opportunity to Work Together

Germany and China already represent one of the world's largest trading partnerships. Whereas Germany has powerful capabilities in fundamental research and upstream technologies (Maine & Gamsey, 2006), China is quick at driving mass adoption of new innovations. Over the years, Germany has accumulated significant knowledge in the field of healthy ageing, bringing welfare to many individuals in its extensive ageing population. Germany's industrial strength and practices of care education, finance and community design have the potential to find many useful applications in China. On the other hand, China possesses world-leading application technologies and digital platforms. Chinese government initiatives encourage large-scale industrial, social and policy experiments. Germany and China may find many overlaps in their respective needs as well as opportunities for the sharing of best practices and knowledge-exchange. This complementarity could form the basis for a mutually beneficial and long-lasting collaboration.

3.2 Trends, Methods and Digital Health Solutions

Throughout the course of human evolution, innovations have repeatedly enhanced living quality by providing solutions to emerging challenges. The current rate of innovation gives cause for confidence in the prospect of successfully addressing a range of growing social challenges such as population ageing. Advances on the internet of things, artificial intelligence and big data will only drive further evolutions in the direction of active living, increased mobility, and the sharing of experiences. Innovations are also increasingly being developed from a multi-disciplinary perspective (Sharp, 2014). Policy makers, researchers and industries are exploring and redefining the boundaries of digital health. New initiatives are interacting with existing systems, and the learning and feedback loops have the potential to inspire fresh rounds of innovation (Bouncken & Fredrich, 2016; Jarvenpaa & Valikangas, 2014).

To facilitate interactions across different sectors, design processes can be utilised to stimulate creativity and unlock the imagination (Brown, 2008). The ageing society is unprecedented. Experts agree that we cannot predict the future changes (Selin et al., 2015). Nevertheless, we can develop plausible future scenarios (Dunne & Raby, 2013) and engage in strategic dialogues across different disciplines. Strategic dialogues of this kind trigger further development effort, more experimentation, and yet more innovative solutions. Speculative design is inclusive of participants and users, while serving to stimulate open discussion, and foster a continuous rather than a closed process.

3.2.1 Merck KGaA, Darmstadt, Germany and ma ma Project: A Case Study of Digital Health and Speculative Design

To encourage strategic discussion among key stakeholders, Merck KGaA, Darmstadt, Germany sponsored a speculative design project that generated potential digital health applications in 2017. The project was conducted collaboratively with ma ma Interactive System Design (ma ma). The initial idea was inspired by innovations in display technology and the ageing society. Digitalisation and display innovations are seamlessly connecting different disciplines and making human–machine interfaces ever more user-friendly. Mobile phones, tablet PCs and wearable devices are providing rapidly evolving interfaces and user experiences. Such innovations have reshaped communication and the society evolves together with these digital applications. Within the project, relevant healthcare, technology and design trends have been studied and analysed.

Relevant Healthcare, Technology and Design Trends

From its analysis of healthcare trends, the research team found that, while burgeoning healthcare spending, partially driven by the ageing society, requires further management, productivity needs to be improved. Similarly, the researchers also detected a pattern whereby increasingly less trained medical staff are being called on to perform more and more complex tasks. Alternative business models, care systems and cost models are currently under development by both policy makers and private sectors, such as patients being offered treatment at home rather than having to stay in hospital overnight. Individuals, family members and communities will be encouraged by policy makers to take more ownership of their own well-being. The 'caregiver economy' is operational, developed and expanded by policy makers and businesses (Connelly et al., 2018; J. Walter Thompson Group 2016) To strengthen and enrich the care education is essential to support the new social needs.

On the other hand, more and more electronic devices are connecting everybody's everyday lives to each other. 'Life-sharing' means sharing other people's lives by communicating experiences and activities through various communication channels. They can be used to reduce the loneliness of the elderly and enhance the sense of belonging. Depending on the individual's needs, different devices, tools and services may be selected to provide appropriate data collection and monitoring to prevent disease and enhance healthy living. Telemedicine has met with high acceptance from both patients and doctors, and further development of the necessary infrastructure is rapidly ongoing. Furthermore, anticipatory diagnosis and prevention will play a more important role in taking care of a healthier and active ageing population. It can also be a key contributor for age-friendly cities. Figure 3.1 demonstrates the relocation of diagnosis from lab to hospital and of treatment from hospital to home.

In the technology domain, the research team has witnessed significant advances in on-site laboratory analysis technologies. These technologies enable healthcare professionals to provide better patient management and faster analysis, and they ensure optimised clinical outcomes. These devices should become easier to use in future, which would also render the devices more robust in terms of storage and usage. For example, lab-on-a-chip technology (e.g. National Technology & Engineering Solutions of Sandia, 2011) allows procedures which previously required a chemical analytical laboratory to be performed on a miniature scale, within a portable or handheld device. This miniaturised technology enhances the convenience of testing and reduces the cost of analysis. By means of these digital tools, healthcare technologies will access big data and make significant further contributions to personalised healthcare, making precision medicine a reality (Frog, 2017).

From the design perspective, the research team focused on the trends of 'calm technology' (Weiser & Brown, 1995), 'invisible technology' (Byford, 2012) and 'inclusive design' (Design Council, 2006). The interaction between a calm technology and its user occurs in the user's periphery, rather than at the centre of the user's attention, which is more intrusive. Invisible technology is embedded in the environment and, as its name suggests, is entirely invisible when it is not in use. These design trends enhance the wellbeing and daily experience of people, especially regarding the elderly as they prefer natural and peaceful surroundings. Inclusive design takes account of the diversity of people in formulating design decisions. As people age, they do have very diversified capabilities, needs and aspirations. The application of inclusive design principle can make buildings, infrastructures, products and services more accessible, self-explanatory and supportive. It can also help reduce demand for qualified professionals or become complementary to care education systems, since users can cover more of their own needs themselves.

Using speculative design and the associated research findings, ma ma developed eight digital health applications that responded to social, healthcare, technology and design trends. Although neither Merck KGaA, Darmstadt, Germany nor ma ma currently intend to develop these products, speculative design plays an important role in encouraging strategic dialogue, unlocking creativity and fostering potential developments. Three of these digital health solutions are discussed below.

Portable Telehealth Kit

This innovation, see Fig. 3.2, helps users access medical support at home as well as when they are out and about. The inspiration sprang from the thought: "What if your family doctor could be available 24/7?" The portable telehealth kit consists of three diagnostic devices equipped with sensors. It can measure physical symptoms such as body temperature, the sound of the heart, lungs and bronchi, and it can perform visual examinations of the nasopharyngeal zone (e.g. throat, shape and colour of the lymph). Furthermore, a flexible display is integrated into the device. This can demonstrate the results of different measurements, assisting the patient during conversations with their family doctor.

In the face of today's ageing populations, this device appears to be an important tool that can enhance the wellbeing of the older societal members who may find it difficult to make regular visits to hospital. Indeed, it should be noted that elderly people often prefer to live in the countryside, and, hence, the long journey to the nearest hospital can be challenging for family members and carers alike. The portable telehealth kit device can, therefore, provide significant benefits to users, family members and the family doctors.

Personal Health Hub

What if your health assistant could be a smart design object? This question has sparked off the invention of the personal health hub. The device measures vital indicators such as heart rate, blood pressure and blood oxygen (SpO2/Pulse Oximetry) and produces an electrocardiogram (ECG). Users initiate measurement by placing both hands on the hub's two golden sensor surfaces. The hub houses a circular multitouch display which can show different types of measurement and other information. If the vital indicators flag a critical condition, a doctor or a medical service centre will be contacted for further assistance. The hub can give health-related recommendations based on predictive analysis health algorithms. When not in use, it functions as an ambient display that decorates its environment, as Fig. 3.3 demonstrates.

A Speculative Design Showcase [Good Example of Sino-German Cooperation]

On 31st January 2018, German Healthcare Partnership, Merck KGaA, Darmstadt, Germany and VFA (*Verband Forschender Arzneimittelhersteller* [Association of Research-Based Pharmaceutical Companies]) hosted a digital health event at Allianz Forum, Berlin. This event, 'Digital Health—Improving Prevention, Prediction and Care: A Global, European and German Opportunity,' welcomed more than 300 guests including policy makers, industrial leaders and academic researchers. The organisers exhibited the Merck-sponsored speculative design prototypes in the venue and invited guests to discuss their responses to them. The installation concept, called the Health Futures Gallery, attracted many event participants. A Chinese enterprise

Fig. 3.3 Personal health hub prototype (*Source* ma ma Interactive System Design 2017)

sponsored the most recent displays. During the day-long event, organisers and audience members engaged in energetic discussions and put forward ideas which were inspired by their experiences with the installation. These conversations sparked future cooperation projects. The event and the Health Futures Gallery also demonstrated the potential for Sino-German collaboration based on the two countries' common interests and complementary natures.

3.2.2 Implications to Care Education, Long-Term Health Insurance and Age-Friendly Cities

This book highlights the importance of care education, long-term health insurance and age-friendly cities for the ageing section of society. In the connected world, systematic changes require co-creation and the emergence of new ecosystems (Jarvenpaa & Valikangas, 2014). The article will elaborate these aspects and discuss the implications with the digital health project.

Care Education

As societies age, demand for care will continue to grow among the elderly and the unwell. Care can be provided at various sites such as hospitals, nursing homes, communities and at home. To support the society transformation, care education needs to respond with higher flexibility, productivity and quality. Consequently, care providers can guarantee sufficient care for rising number of elderly people and boost the treatment capabilities with the growing demand. Otherwise, the system can quickly become unsustainable.

Technologies can also make care provision more flexible and accessible. Digital health has the potential to enhance the quality of life of the elderly, boost the output of carers, and improve the wellbeing of families. Digital tools can help users monitor their health and communicate with carers. Assistive robotics have shown more and more potential for use in both professional and domestic environments. Furthermore, new applications have the potential to encourage the elderly, family members and other potential carers (e.g. less trained staff, social workers) to take an increased share of the responsibility for more independence and better quality of life.

Care education should include these latest developments while the regulatory environment will need to be adapted to assistive care technologies, real-world evidence and personalised healthcare. Tomorrow's carers will be more familiar with the latest available digital tools and will be able to use them in clinics and even remotely. The interactions with various stakeholders will be much more collaborative. It is encouraging that educational institutions, healthcare policy makers, care providers and technology companies in Germany and in China alike are cooperating closely to provide ever better solutions.

Long-term Health Insurance

The world's swiftly ageing population is exerting ever more strains on the health insurance industry (Financial Times, 2014). Average lifespan has increased significantly, and most pensioners now live much longer than the age of 65. Therefore, as many people will likely endure episodes of ill health, especially in the later stages of life, they will require adequate insurance cover (Rhee et al., 2015). These factors tug more and more on the strings of the public purse and mount pressure on private insurance schemes. Historical data is no longer sufficient to predict the future, and thus current working models no longer appear sustainable.

Insurance providers are eagerly endeavouring to leverage new technologies, such as digital health devices, and access the latest sophisticated big data to assess life expectancies and associated risk profiles. More and more real-world evidence provided by new technologies is fuelling these developments and driving regulatory changes. Insurance providers are building up new capabilities to assess, analyse and implement the wealth of data available. Algorithms and machine learning are embedded into these new mechanisms. Trueman et al. (2010) describe various emerging schemes.

At the same time, policymakers need to evaluate current regulations and to adapt them in accordance with these new societal developments The public will soon realise the true extent of the challenges posed by the increased dependence ratio within society: consequences on elderly people's healthcare insurance policies, extra demands on younger members of the labour force (especially if they wish to benefit from pensions in later life) and impacts on economic growth.

German policymakers have raised the retirement age to tackle this challenge (Hess, 2017). Thanks to its strong economy, Germany has attracted an influx of young workers, and pressures on the country's social system have been somewhat alleviated. However, other European countries are facing the prospect of social unrest in response to the necessary reforms. Experiences gained in Germany may provide helpful insights for Europe and for China, which will face similar challenges in the coming years. Meanwhile, China is among the leaders in the world at collecting, analysing and applying big data to industry and social policy making. It is highly likely that the Chinese systems will be able to provide new insights for Germany too. Nevertheless, China and Germany have currently very different data security standards and regulations and both will review these standards extensively.

Age-friendly Cities

In response to developments driven by demographic changes and urbanisation, many policymakers around the world are developing age-friendly cities and communities. Cities must further encourage the principles of wellbeing, social inclusion and security as people age (WHO, 2007). In the era of open innovation, cities are becoming living innovation laboratories for public and private sectors and citizens. (Cohen et al., 2016). Speculative design method can provide inspiring opportunities for continuous discussions, collaborations and experimentations.

Landscaping, buildings and transport systems can significantly shape older people's confidence about their own mobility, thereby potentially encouraging social inclusion and healthy behaviours. The elderly prefer to spend a considerable of time in public parks and green spaces (Yung et al., 2017). Care facilities should be easily accessible via public transport or other means (Wong et al., 2018). Calm design and inclusive design can provide significant benefits for citizens. Digital health solutions are complementary facilitating the transformation.

Traditionally, the automotive industry has a strong impact to the urban life. The demand for age-friendly cities and environmental sustainability is pushing the industry to undergo significant changes. Electric and fuel-cell cars have vastly expanded across China. This trend helps to reduce pollution and grants China the chance to take the lead in future developments (Kuang et al., 2018). Autonomous driving technologies are also currently undergoing rapid development. It has the potential to greatly ease mobility for the elderly. Furthermore, car-sharing can sometimes allow those of the older generation to participate more actively in society and can thus enhance inclusion. New services, platforms and business models will increasingly meet the needs of users as they age and will enrich the urban experience, boost convenience and lessen the risk of social isolation. While China is rapidly expanding in the fields of electric vehicles, the shared economy and new business models, German carmakers are also trying to maintain their traditionally premier position, strengthen the infrastructure and capacities and acquiring new expertise (Regan, 2017). Germany and China will collaborate, compete and co-innovate in these new mobility sectors for future urban lives.

3.3 Summary

Population ageing is a global phenomenon. Germany and China are both experiencing significant technological, economic social and policy changes that will only accelerate in the coming years. Despite the contextual differences between them, Germany and China possess complementary expertise and infrastructures, and collaboration between the two countries has the potential to foster substantial win–win scenarios for both.

Throughout the course of human evolution, innovations have repeatedly enhanced the quality of life by providing solutions to emerging challenges. The applications of digital health can potentially provide significant contributions to the new social norm. The use of speculative design can stimulate continuous cross-discipline innovation. Similar approaches may be applied on an experimental basis in areas such as care education, long-term health insurance and urban living.

To build up a successful ecosystem for the future, the author encourages more collaboration and experimentation across industries, more private–public partnerships, and more knowledge exchange between countries. Germany and China boast a track record of successful partnership which spans over the past decades. The challenges and opportunities of a globally ageing population will further unite both countries in the search for innovation, prosperity and an improved quality of life.

References

Bouncken R, Fredrich V (2016) Joint knowledge creation and protection in competitive business model innovation. In *Academy of Management Annual Meeting Proceedings*, January 01.

Brown, T. (2008). DesignThinking. *Harvard Business Review, 86*(7), 84- - 92, 141.

Byford, S. (2012). Good design is invisible: an interview with iA's Oliver Reichenstein. *The Verge (24.07.2012).* Retrieved from: https://www.theverge.com/2012/7/24/3177332/ia-oliver-reichenstein-writer-interview-good-design-is-invisible. January 06, 2018.

Cohen, B. E., Almirall, E., & Chesbrough, H. (2016). The city as a lab: Open innovation meets the collaborative economy. *California Management Review, 59*(1), 5–13. https://doi.org/10.1177/0008125616683951

Connelly, R., Dong, X., Jacobsen, J., & Zhao, Y. (2018). The care economy in post-reform China: Feminist research on unpaid and paid work and well-being. *Feminist Economics, 24*(2), 1–30. https://doi.org/10.1080/13545701.2018.1441534

Design Council. (2006). *The principles of inclusive design.* Retrieved from: https://www.designcouncil.org.uk/resources/guide/principles-inclusive-design. December 12, 2017.

Dunne, A., & Raby, F. (2013). *Speculative everything: Design, fiction, and social dreaming.* MIT Press.

Financial Times. (2014). *The Silver Economy: Life insurers take strain of ageing population (27.10.2014).* Retrieved from https://www.ft.com/content/f9b1e57a-32a9-11e4-a5a2-00144feabdc0. December 12, 2017.

Frog. (2017). *Techtrends 2017.* Retrieved from https://www.frogdesign.com/techtrends2017. December 12, 2017.

German Healthcare Partnership. (2018). *Digital health—Improving prevention, prediction and care. A global, European and German opportunity.* Retrieved from http://ghp-initiative.de/digital-health-improving-prevention-prediction-and-care-a-global-european-and-german-opportunity-31st-jan-2018/. February 02, 2018

Hess, M. (2017). Rising preferred retirement age in Europe: Are Europe's future pensioners adapting to pension system reforms? *Journal of Ageing and Social Policy, 29*(3), 245–261. https://doi.org/10.1080/08959420.2016.1255082

J. Walter Thompson Intelligence. (2016). *The future 100, trends and change to watch in 2017*. Retrieved from https://www.jwtintelligence.com/2016/12/future-100-trends-change-2017/. December 12, 2017.

Jarvenpaa, S., & Välikangas, S. (2014). Opportunity creation in innovation networks: Interactive revealing practices. *California Management Review, 57*(1), 67–87. https://doi.org/10.1525/cmr.2014.57.1.67

Kuang, X., Zhao, F., Liu, H. H., & Liu, Z. (2018). Intelligent connected vehicles: The industrial practices and impacts on automotive value-chains in China. *Asia Pacific Business Review, 24*(1), 1–21. https://doi.org/10.1080/13602381.2017.1340178

Maine, E., & Garnsey, E. (2006). Commercialising generic technology: The case of advanced materials ventures. *Research Policy, 35*(3), 375–393. https://doi.org/10.2139/ssrn.1923117

National Technology and Engineering Solutions of Sandia. (2011). *Rapid automated point-of-care. System (RapiDx)*. Retrieved from https://ip.sandia.gov/technology.do/techID=81. January 13, 2018.

Rapoza, K. (2017). China's ageing population becoming more of a problem. *Forbes*. Retrieved from https://www.forbes.com/sites/kenrapoza/2017/02/21/chinas-ageing-population-becoming-more-of-a-problem/#28473b0e140f. January 05, 2018.

Regan, J. (2017). *German carmakers urged to challenge Tesla by senior Merkel aide*. Retrieved from https://www.bloomberg.com/news/articles/2017-08-03/germany-giving-gigafactory-a-home-in-latest-challenge-to-tesla. January 05, 2018.

Rhee, J. C., Done, N., & Anderson, G. F. (2015). Considering long-term care insurance for middle-income countries: Comparing South Korea with Japan and Germany. *Health Policy, 119*(10), 1319–1329. https://doi.org/10.1016/j.healthpol.2015.06.001

Selin, C., Kimbell, L., Ramirez, R., & Bhatti, Y. (2015). Scenarios and design: Scoping the dialogue space. *Futures, 74*, 4–17. https://doi.org/10.1016/j.futures.2015.06.002

Sharp, P. (2014). Meeting global challenges: Discovery and innovation through convergence. *Science, 346*(6216), 1468–1471. https://doi.org/10.1126/science.aaa3192

Tan, X., Wu, Q., & Shao, H. (2018). Global commitment and China's endeavours to promote health and achieve sustainable development goals. *Journal of Health, Population and Nutrition, 37*(8), 37–38. https://doi.org/10.1186/s41043-018-0139-z

Trueman, P., Grainger, D. L., & Downs, K. E. (2010). Coverage with evidence development: Applications and issues. *International Journal of Technology Assessment in Health Care, 26*(1), 79–85. https://doi.org/10.1017/S0266462309990882

Weiser, M., & Brown, J. S. (1995). *Designing calm technology*. Retrieved from http://citeseerx.ist.psu.edu/viewdoc/download;jsessionid=EC862F617ECDAF2C89ED9E95C7E45A6B?doi=10.1.1.123.8091&rep=rep1&type=pdf. January 06, 2018.

WHO, World Health Organisation. (2007). *Global age-friendly cities: A guide*. Retrieved from: http://www.who.int/ageing/publications/Global_age_friendly_cities_Guide_English.pdf. January 03, 2018.

WHO, World Health Organisation. (2015). *World report on ageing and health*. Retrieved from http://www.who.int/ageing/events/world-report-2015-launch/en/. January 03, 2018.

Wong, R. C. P., Szeto, W. Y., Yang, L., Li, Y. C., & Wong, S. C. (2018). Public transport policy measures for improving elderly mobility. *Transport Policy, 63*, 73–79. https://doi.org/10.1016/j.tranpol.2017.12.015

Yung, E., Ho, W., & Chan, E. (2017). Elderly satisfaction with planning and design of public parks in high density old districts: An ordered logit model. *Landscape and Urban Planning, 165*, 39–53. https://doi.org/10.1016/j.landurbplan.2017.05.006

Luc Yao is responsible for several innovation and transformation projects for Merck KGaA. Luc verifies market potential, constructs ecosystems across value chains, and fosters technological innovations through effective commercialisation. Since 1668 Merck has been in the forefront of science and technology discovery. Many of these brilliant ideas spanning the healthcare, life science and specialty materi-als fields, continue to provide profound benefits for our societies and touch our lives in diverse ways. Luc Yao is an academic visitor at Oxford Institute of Popu-lation Ageing, and a speaker for Digital Health at the German Healthcare Partner-ship. Luc Yao obtained a Master of Business Administration from Manchester Business School, UK, and a Master of Science for Consulting and Coaching for Change from HEC Paris and Said Business School, Oxford University. Luc is based in Darmstadt Germany and is active in the electronics industry, entrepre-neurships, and the Open Innovation networks.

Part II
Care Insurance

Chapter 4
Social Long-Term Care Insurance in Germany

Steffen Fleßa

4.1 Introduction

The Federal Republic of Germany has a long-term experience in providing social protection to its population. In 1883, Reichskanzler Bismarck launched the Social Health Insurance in Germany as the first compulsory social protection system for the formal sector in this country. Since then, coverage has increased step-wise (Bärnighausen & Sauerborn, 2002). Important milestones were the work accident insurance (1884), invalidity insurance and old-age pension fund (1889), Pension Fund for White Collar Workers (1913), and the Unemployment fund and re-employment act (1927). After several reforms of these social insurances the re-united Germany finalised its efforts to provide social protection to its population by launching the social long-term care (LTC) insurance in 1995 and opened chapter XI of the Social Law. Its purpose is the provision of social protection against the financial risk of long-term nursing care that goes beyond the coverage of the health insurance. It is the so-called fifth pillar of the social protection system (Kronenberg, 2015) and is compulsory for every wage-earner in Germany either within the Social LTC Insurance Fund (year 2016: 72 million members) or a Private LTC Insurance (year 2016: 9 million members) (BMG 2018). The other pillars are health insurance (I), work injury insurance (II), pension insurance (III) and unemployment insurance (IV).

In the year 2016, some 2.9 million citizens benefited from the insurance. The majority of them were old and required assistance in their activities of daily living (ADL), such as dressing, bathing, eating, toileting, continence, and mobility (Beckers & Buck, 2014). The total expenditure of the LTC insurance was some 31 billion Euro. Thus, within a few years the fifth pillar has become a crucial component of high relevance for the population and the entire social protection system.

S. Fleßa (✉)
University of Greifswald, Greifswald, Germany
e-mail: steffen.flessa@uni-greifswald.de

© The Author(s) 2025
Deutsche Gesellschaft für Internationale Zusammenarbeit (GIZ) GmbH et al., (eds.),
Sustainable Aging, https://doi.org/10.1007/978-3-662-69139-7_4

In this paper we would like to present the foundations of LTC insurance in Germany. For this purpose, we will first point out different options to develop such a system to understand the German way. We do this in the knowledge that the German system is merely one alternative of many. It is based on its historical pathway and not a blueprint for any other nation. It is also under steady reform. However, knowing the options and the pathway chosen by Germany might be encouraging for other nations facing the same problems that induced the launching of this insurance in Germany in the last century. Afterwards, we will go into details of the German system and the current state of its development. The paper closes with some challenges calling for a steady process of reforms.

Long-term care insurances provide protection against the financial risk of requiring long-term assistance for activities of daily living not covered by health insurances.

4.2 Options

4.2.1 Benefits

A health insurance protects against the financial risks of sickness, i.e., medical expenditure such as charges of physicians, hospital care, medication, physiotherapy etc. (Fleßa & Greiner, 2013). In particular in hospitals, this does not only include the costs of medical and para-medical professionals, but also of nursing care during the period of diagnosis, treatment and healing. However, health insurances do not cover the costs of LTC necessary to perform the activities of daily living. Frequently, old or disabled people require assistance to get washed, eat, go to the toilet, communicate and walk not only while they are treated for a specific disease, but for a very long time—maybe life-long. The respective costs are not covered by the health insurance as they are not medical costs for diagnosis or treatment, but costs to perform the activities of daily living. Still, these costs can be substantial. Consequently, LTC insurances cover the financial risk of costs of assisting in the daily living of people (Engelman, 2008). This can include:

- **Care at home (homecare): This care can be offered by professional (ambulatory) nurses or by volunteers, such as relatives.**
- **Tools and equipment for nursing care at home: Regularly, patients requiring LTC also need special devices, such as nursing beds, wheelchairs etc.**

- **Partial inpatient nursing care: Patient requiring long-term nursing care might stay at home but require inpatient[1] services for a limited time only, such as after discharge from hospital, during day or night or while relatives are on leave.**
- **Nursing care in nursing homes (nursing inpatient care): Some patients require nursing care in homes where they stay for a long time. The respective costs also include accommodation and food.**

Any long-term health insurance has to decide what it wants to cover. In Germany, homecare, equipment, partial inpatient and inpatient nursing care are covered. However, costs of accommodation and food are not covered as it is assumed that these costs will occur whether the patient needs nursing care or not. Furthermore, any LTC insurance has to decide what "long-term" actually means. In Germany, the LTC insurance is relevant if the necessity of care exceeds six months.

German long-term care insurance covers home care, care equipment, partial inpatient and nursing inpatient care, but not the costs of accommodation and food.

4.2.2 Private and Social Insurance

Long-term care insurances can be organised as private or social funds just as health insurances can be private or social. In Germany, every employee with a salary of less than 59,400 Euro p.a. is obliged to be member of one of some 110 social health insurance funds (§ 6 Abs. 6 SGB V). This amount is adjusted every year. Every self-employed and every employee earning more than this amount can chose to be privately insured. Currently, some 89% of the population is insured in a social health insurance fund. This includes their families as these funds offer a premium-free family insurance (Fleßa & Greiner, 2013).

With the introduction of the LTC insurance it was decided to apply the same rules (Richter, 2017). Everybody who is obliged to be member of a social health insurance fund will also be in the respective social LTC fund. All others can choose to become voluntary members of a social LTC insurance fund or become members of a private insurance to cover the LTC risk. Thus, the entire population is obliged to be insured. The term "voluntary" has two meanings for LTC insurances: It can

---

[1] The term "inpatient" is frequently used to describe a person requiring long-term care in a special care institution, such as a home of the elderly. It is misleading as these people are not patients in the traditional meaning. They are clients and residents of an institution offering care. However, we will use this term in accordance with the literature.

mean that somebody chooses to stay in the social LTC insurance fund although his income is sufficiently high that he could go for the private fund (which usually has lower premiums for such high incomes). Secondly, it can mean that somebody buys additional coverage beyond the normal LTC insurance, for instance for a single room in nursing homes.

> 89% of the population is member of a social long-term care insurance fund. Only self-employed and people with high income can choose the private long-term care insurance.

4.2.3 Financing

Financing of health care and LTC differs strongly between social and private insurance. We present the difference between pay-as-you-go and capital cover system. Afterwards we will focus on the alternatives of financing the social LTC insurance as it covers 89% of the German population.

4.2.4 Pay-As-You-Go Versus Capital Cover System

The principle of insurance is that the risk of a catastrophic financial loss caused by a long-term necessity of care is pooled and shouldered by many people (insurance pool). The insurance has to recover its expenditure by its income (principle of equivalence) which can be achieved in two ways (Zweifel et al., 2009). Firstly, the total of all premium payments of one insurance from the first day to the last day of his insurance contract covers on average all expenditure of this insured person. As most people will not require LTC while they are young, they accumulate funds with the insurance which will be utilised when they are old and require LTC. At the end of life, the accumulated fund has been consumed. As the company insured many members, it suffices if the total life-long average expenditure of an insured person is equivalent to the total average life-long income from this person. This is called capital cover system (Kapitaldeckungsverfahren) or individual equivalence (Stiglitz & Rosengard, 2015).

Figure 4.1 exhibits the principle. The area between the beginning of the insurance contract and the intercept between the two curves must be equal to or greater to the area between the intercept and the end of the insurance contract. However, the capital fund will earn interest which is not reflected in this chart.

Social LTC insurances do not follow this principle, i.e., they do not build-up a capital fund to cover future expenditure for LTC. Instead, the equivalence principle

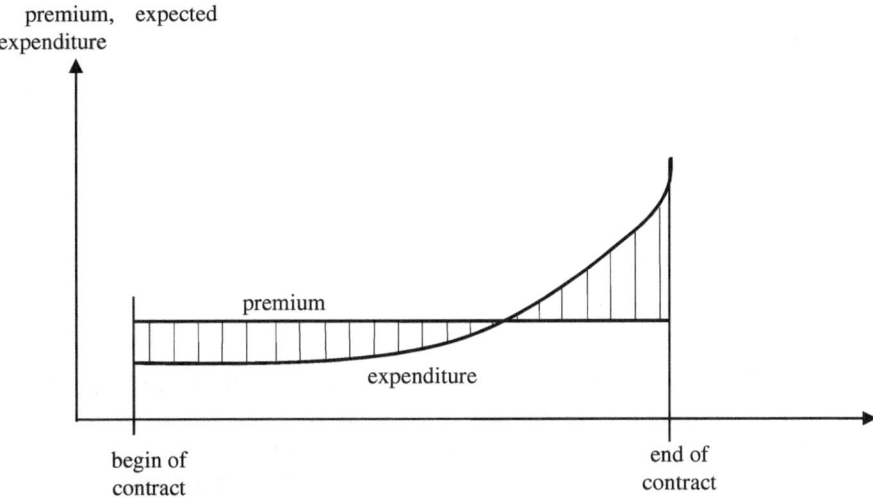

Fig. 4.1 Individual equivalence principle (*Source* Fleßa & Greiner, 2013)

requires that the total expenditure for LTC in a certain year must be identical with the total of all premium contributions. This is called pay-as-you-go (Umlageverfahren) or collective equivalence (Johansson, 2000; Zweifel et al., 2009).

Private LTC insurances use the capital cover principle while statutory insurances apply the pay-as-you-go principle. Other differences of both types are the methodology of premium calculation, the relevance of individual risk and the role of the interest rate. The premium of a private insurance is proportional to the risk, i.e., the higher the likelihood of requiring LTC the higher is the premium. For social funds the premium depends on the income of the insured, i.e., the higher the income the higher the premium, i.e., the social fund is based on the principle of solidarity: old people pay the same premium as young people, men as women, the healthy as the sick. Consequently, people with a very high income but a low risk (e.g. young and healthy professionals with high income) tend to prefer the private LTC insurance.

The big disadvantage of the collective equivalence principle is that its sustainability strongly depends on the demographic structure. An ageing population will induce a financial problem as less young and healthy people have to cover the LTC costs of older and sicker people. This is no problem for the private insurer as the older people either pay higher premiums or have entered the insurance long ago so that they could accumulate funds to cover their costs.

Some 89% of the German population is member of a social LTC insurance or is an insured family member. The guiding principles for them are the pay-as-you-go, income dependency of premium and solidarity.

Long-term care insurance follows the principle of collective equivalence, i.e., the total expenditure for long-term care of a certain year must balance the total premium collection of the same year. The premium depends on the income of the insured, not their individual risk.

4.2.5   Sources of Funding

Another important financing decision is the source of financing for LTC expenditure. Funds can come from the Government, the employer, employees or charities. The German social LTC insurance does not receive any funds from the Government or charities. The strict distinction between Government and insurance fund is a heritage of the Bismarck system relevant for all social protection mechanisms in Germany. The Government neither finances the funds nor interferes in its management. Instead, the social insurance funds are managed and financed by employers and employees (Fischer, 2015).

In the year 2018, every member pays 2.55% of his income to the LTC insurance fund with a maximum of 1354.05 Euro p.a. (2.55% of 53,100 Euro). This rate applies for pensioners (born before 1.1.1940), youths (< 23 years), unemployed and soldiers (Sozialversicherung kompetent, 2018). Any members without children pay 2.8%. The employer pays 1.275% of this amount.

- Example A: Employee, 1 child, 25,000 Euro annual income.
- Employer: 1.275% of 25,000 Euro = 318.75 Euro p.a.
- Employee: 1.275% of 25,000 Euro = 318.75 Euro p.a.
- Example B: Employee, no children, 25,000 Euro annual income
- Employer: 1.275% of 25,000 Euro = 318.75 Euro p.a.
- Employee: 1.525% of 25,000 Euro = 381.25 Euro p.a.
- Example C: Employee, 1 child, 60,000 Euro annual income.
- Employer: 1.275% of 53,100 Euro = 667.03 Euro p.a.
- Employee: 1.2575% of 53,100 Euro = 667.03 Euro p.a.
- Example D: Employee, no children, 60,000 Euro annual income
- Employer: 1.257% of 53,100 Euro = 667.03 Euro p.a.
- Employee: 1.525% of 53,100 Euro = 809.78 Euro p.a.
- Example E: Self-employed, 1 child, 45,000 Euro annual income
- Self-employed: 2.55% of 45,000 Euro = 1147.50 Euro p.a.

The rates for the state of Saxony (employer: 0.775%; employee: 1.775%) and for civil servants (employer: 0%; employee: 1.275%) differ slightly from these standards.

There was some discussion whether it is fair that people without children pay more to the fund than parents. Some argue that being without child is frequently not a personal choice so that it is unfair to punish them for this condition with higher rates. Others argue that parents already contribute to the sustainability of the social system by having children who will pay for the elderly and take care of their parents when they are adult. In addition, raising children is quite expensive so that it seems fair to release parents from some financial burden.

> The German social long-term care insurance is fully financed by employers and employees without any Government contribution. For most members, employer and employee pay the same amount to the fund.

4.2.6   Pooling of Risks

Social health insurance and social LTC insurance are rather similar concepts so that it could be possible to unite the funds in one risk-sharing pool. Generally, the greater the risk pool, the less is the risk for the insurer. This calls for a unification of risk pools between the two types of insurance. However, the German Government decided to separate the management and risk pooling of both insurances as health and LTC are different categories to avoid cross-subsidies between these separate pools. It must not happen that funds from the LTC pool are used to cover deficits of the health insurance pool or vice versa. Consequently, pooling of risks and of management are separated.

However, it is an ongoing discussion whether this was a proper decision as it has several disadvantages. Firstly, separate risk pools are smaller and have higher variation of expenditure than one united pool. This disadvantage is not very big as the pools are large enough even while separate. Secondly, some health care providers have to deal with two different financers and issue separate bills for the same patient as some services have to be paid by the health insurance (medical treatment) and others by the LTC insurance. Although this might seem cumbersome, in reality it is not really challenging. The main problem of separating health and LTC insurance is, thirdly, the fact that the two insurances might "play games", i.e., each of them might point at the other insurance type for covering a certain service. At the end, the patient might not receive what he requires because the health insurance is sure that the LTC insurance has to pay for it while the latter is convinced that the health insurance is responsible.

Health insurance and long-term care insurance are two separate funds and risk
pools.

4.2.7 Payment Mechanisms

In principle, services can be paid by different mechanisms:

- Flat rates: a certain amount is paid for a defined result. For instance, hospital
  services are paid on a flat rate basis called "diagnosis related group". The
  payer does not care how long the patient resides in hospital or what specific
  services he receives. The flat rate is a price for the treatment (Fleßa, 2018).
  A flat rate in LTC means that a fixed amount is paid to the patient or a
  caregiver if a patient requires LTC irrespective of the number and category
  of services rendered. Flat rates incorporate the risk that the quantity and
  quality of services are low.
- Daily Rates: a certain amount is paid per day or night which recovers all the
  costs, i.e., no additional amount is paid if the client receives more services
  during that day. Usually, the amount paid is identical for each day. Daily
  rates have the risk that the number of days are unnecessarily high.
- Fee-for-Service: each individual service is paid for. Consequently, there is
  a risk that the number of high-cost services is increased unnecessarily.

Most services paid for by the LTC insurance are daily rates or compensation for
payment (e.g., equipment). For instance, people admitted to a home of the elderly
receive a certain amount for each day they spend in this home to pay for the nursing
care cost. It is an amount per month, but in the month of admission and of discharge/
death the amount is proportional to the length of stay. Homecare is frequently also
a lump sum paid to the relatives with which they can pay the ambulatory nursing
services. If they decide to give the care themselves, they can keep the respective
lump sum (which is generally 50% of the lump sum paid for professional care
givers). Equipment, such as wheelchairs or nursing beds, are usually paid by the
LTC insurance per piece up to a certain maximum defined in the regulations.

The dependency on care will differ between beneficiaries reaching from minor
assistance in daily medication (e.g., insulin injection) to full-time professional
nursing care. Thus, the payment mechanism has to develop a system of assessing the
need of care and categorising beneficiaries into classes, levels or grades of care.

The German social long-term care insurance pays daily rates for nursing costs or lump-sums for equipment to the patient or his relatives. Flat rates, such as the hospital DRGs, are not (yet) used in long-term nursing care.

Summarising, we can state that an LTC insurance has many options of designing its system. The German system is based on 135 years of tradition of a social insurance deeply grounded in the principles of solidarity. It is one system that works quite well for Germany, and which holds relevant learnings for other countries. However, it is not a blueprint to copy (Campbell et al., 2010).

4.3 Processes of the German Long-Term Care Insurance

Social Law XI covers the social LTC insurance in Germany. It is supported by several national and sub-national ordinances and guidelines. The system has become so complex that most potential beneficiaries or their relatives need professional help even to apply to become registered as a candidate for the social LTC insurance. In the following we will point at some administrative issues on the pathway towards receiving benefits from the insurance.

4.3.1 Application

The social LTC insurance will only give benefits to its members if they apply for it (§ 33 SGB XI). The insured themselves can apply or their representatives (e.g., for children < 15 yrs. the parents can apply, for elderly only an official legal representative). The social LTC insurance has to decide on the application within five weeks and respond in writing. Otherwise, the insurance has to compensate the applicant irrespective of the assessment result.

In many cases, the level of care has to be increased from time to time, but the respective up-grading process will start only on application. In particular the elderly are sometimes not aware of the complicated procedures and require assistance even for the application. For that purpose, special nursing offices ("Pflegestützpunkt") have been established to advise and support the application process. However, it is not safeguarded that everybody entitled to LTC will apply and finally receive the appropriate financial compensation.

4.3.2 Assessment

In Germany, each individual case is assessed by professional nurses and/or physicians employed by the so-called "Medical Services of the Health Insurances" (Medizinischer Dienst der Krankenkassen, MDK) (Hinselmann, 2016). The name is misleading. Firstly, MDK does not only care about medical conditions, but a major element of its workload is the assessment of individual needs of long-term care clients. Secondly, MDK does not only deal with health insurance but is directly responsible for assessing the members of the social LTC insurances.

Until 31st December 2016 there were three levels of care (I-III) and a special category for patients who are extremely dependent on care. As Table 4.1 shows, the categorisation depended mainly on the time required for proper nursing care.

In 2008, Grade 0 was added as it was realised that some people might have a limited competence in activities of daily living even before they qualify for level I. This is in particular true for people living with dementia and for mentally handicapped. They need constant support of relatives and professional support, e.g., gerontopsychiatric services. Based on the assessment of Level 0 they can get professional advice and a lump sum for professional services.

Table 4.1 Levels of care (until 31st December 2016, *Source* Fleßa)

Care level	I	II	III	Hardship Cases
Minimum demand of care (min/d)	> 90 Min	> 180 Min	> 300 Min	> 420 Min
Conditions	Requiring support for at least two activities of daily living (personal hygiene, nutrition, mobility) at least once a day	Requiring support for at least two activities of daily living (personal hygiene, nutrition, mobility) at least three times a day at different times	Requiring constant assistance for activities of daily living, also at night	Requiring constant assistance for activities of daily living, also at least 120 min at night
Max. time share of domestic assistance	44 min	60 min	60 min	60 min
Potential care givers	Professionals and lay persons	Professionals and lay persons	Professionals and lay persons	Only professionals

In 2017 the system changed strongly (BMG, 2018b). Instead of having the old care levels, we have care grades now. Most beneficiaries were just transferred to the new grades according to the following key:

- **Level 0: → Grade 1**
- **Level I → Grade 2**
- **Level I + reduced competence of daily living → Grade 3**
- **Level II → Grade 3**
- **Level II + reduced competence of daily living → Grade 4**
- **Level III → Grade 4**
- **Level III + reduced competence of daily living → Grade 5**
- **Hardship case → Grade 5**

The new grade does not depend only on the time required to take care of the beneficiary, but professionals of the social LTC insurance assess the limitations and the need of care in different modules (mobility, cognitive and communicative abilities, behaviour and psychological problems, domestic independence, autonomy in the field of medication and treatment, daily living and social contacts). For each module several criteria have been defined, and the total score decides on the grade. The new assessment system is much fairer but also quite time consuming.

The assessors of the MDK analyse the documentation, visit the potential beneficiary and make a structured assessment according to the standards to determine the need and the required costs for care. The assessors will visit the potential beneficiary personally in his home, ask questions and get an impression on his ability in the different dimensions of activities of daily living. The latest version of care grades distinguishes smaller limitations of autonomy (grade 1: 12.5–26.9 points), stronger limitations of autonomy (grade 2: 27–47.4 points), strong limitations of autonomy (grade 3: 47.5–69.9 points), strongest limitations of autonomy (grade 4: 70–89.9 points) and strongest limitations of autonomy with special demands for nursing care (90–100 points). However, the assessor does not only decide on the score, but also gives advice how to improve the situation and where to get professional support (BMG 2018b).

4.3.3 Benefits

It can be quite complicated to understand the set of benefits the beneficiary and his relatives are entitled to. Most important is the distinction between ambulatory home care and inpatient nursing care. Ambulatory home care has several benefits:

1. **Care Allowance:** Beneficiaries receive a monthly cash benefit for privately organised caregivers. Usually, these are family members currently not employed, but also other voluntaries or even hired personnel is possible. As shown in Table 4.2, the amount depends on the level or grade of care currently ranging from 316 € (grade 2) to 901 € (grade 5). The amount is paid as a lump sum irrespective of real costs, but the MDK checks the quality of non-professional care regularly.

2. **Benefit in kind:** The beneficiary or his relatives can select a professional ambulatory nursing service for professional care. The care giver must be registered with the social LTC insurance and bills directly addressed to the insurance. The social LTC insurance subsidises the costs of these services up to a maximum depending on the care grade. Currently, the monthly rate ranges between 689 € (grade 2) and 1,995 € (grade 5).

3. **Technical nursing aids:** If a beneficiary requires a wheelchair, nursing bed or other equipment, the costs are fully covered by the social LTC insurance, but they have to be prescribed by a physician. Relatives do also have the right to attend special nursing courses paid for by the insurance.

4. **Adjustment of living conditions:** In some cases, doors, bathrooms or staircases have to be adjusted. This can be paid for up to a maximum amount per intervention.

5. **Compensation care:** If a relative cannot take care of the beneficiary (e.g., because he is sick himself), a professional service can be hired to compensate him for up to 6 weeks per year. Depending on the care grade this is financed by the social LTC insurance.

6. **Short-term care:** different alternatives of short-term care (e.g., day care, night care, care during holidays of voluntary care givers) etc. are compensated by the social LTC insurance.

7. **Other services:** several specialities, such as care for patients in special living groups (e.g., residential community of people living with dementia) can be financed.

8. **Social protection of voluntary caregivers:** Voluntary care givers receive social protection, e.g., accident insurance, unemployment insurance and health insurance during their service for the beneficiary.

Inpatient care covers the cost of nursing care in a nursing home. It has to be assessed whether the beneficiary cannot continue living at home anymore, either because he has no relatives who can take care of him or because the need of care is so high that it requires full-time professional nursing care. Depending on the care grade up to 2005 € are paid for per month. However, these compensations are regularly not sufficient to pay for anything else but the pure nursing care costs, i.e., rent for the room and payment for food is not included in the calculation.

Table 4.2 gives an overview of the current portfolio of the social LTC insurance (30th June 2017).

Again, it has to be stated that the portfolio of potential services and compensations is wide, but it is very difficult for most beneficiaries and their relatives to understand the system and get appropriate compensation for their costs of long-term care. Regulations and processes are quite complex and require professional advice—not everybody will be able to get what he is entitled to.

> Beneficiaries of the German LTC insurance are classified into five different grades based on a standardised assessment tool. The process starts with the application of the potential beneficiary. A wide range of benefits is granted for professional and voluntary home care as well as for care in homes of the elderly. The actual amount paid depends on the living situation and the care grade of the beneficiary.

Table 4.2 Benefits of social LTC insurance [€] (selected, *Source* Fleßa)

	Grade 1	Grade 2	Grade 3	Grade 4	Grade 5
Cash allowance per month	–	316	545	728	901
Benefit in kind per month	–	689	1.298	1.612	1.995
Compensation care up to 6 weeks p.a. up to € per year		474	817,50	1.092	1.351,50
Short-term care up to 8 weeks p.a. up to € per year:	–	1.612	1.612	1.612	1.612
Part-time institutional care (day and night) up to € monthly	–	689	1.298	1.612	1.995
Compensation for ambulatory nursing up to € per month	125	125	125	125	125
Additional payments for ambulatory services in living groups up to € per month	214	214	214	214	214
Full inpatient nursing care lump sum of € per month	125	770	1.262	1.775	2.005
Technical nursing aids per month	100% of cost				
Improvement of living up to € (per measure)	4000 €				

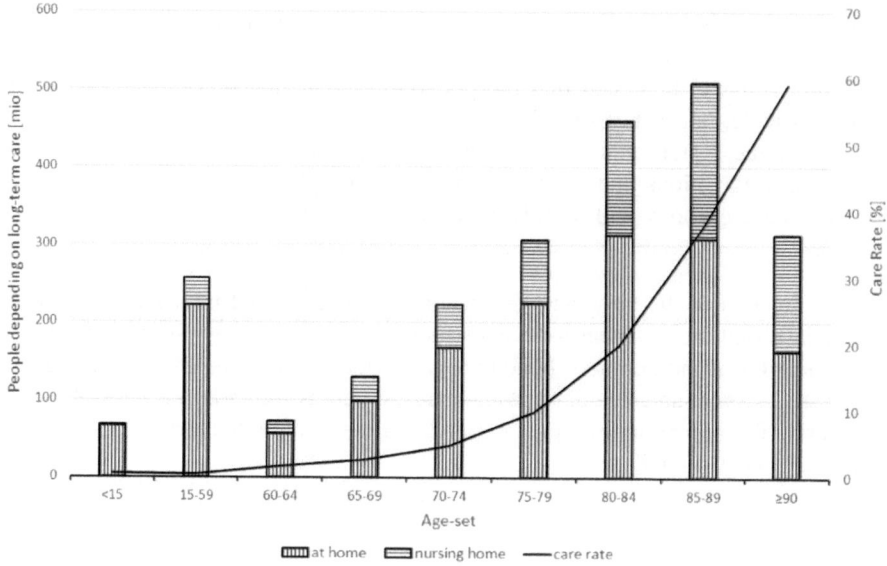

Fig. 4.2 Nursing care and age (*Source* own chart based on data from BMG 2018c)

4.4   Current Situation

4.4.1   Rationale

The main reason of introducing the compulsory LTC insurance in Germany was the steadily growing number of older people requiring that kind of care. Although nursing care is also relevant for younger people and even for children, there is a clear correlation between age and the need for LTC. Figure 4.2 underlines this finding. While the rate of people depending on nursing care is less than 2% in the age-set from 60 to 65, it is over-proportionally growing to up to 60% in the age-set of people older than 90 years. Although the growth might not be linear (i.e., as more and more people will become dependent on nursing care when they get even older) it is obvious that an ageing population will require more LTC. It is also clear that technical progress in medicine, hygiene, public protection etc. will increase the life expectancy even more. The German Statistical Office points out that in the year 2005 some 16 million Germans were 65 years or above, while it estimates that the number will increase to 23 million by the year 2050 (Eisenmenger et al., 2006).

The ageing population will have a tremendous impact on the prevalence of several conditions, in particular chronic-degenerative diseases. With a wide range of scenarios, it is estimated that the number of the elderly suffering from dementia will increase from 935,000 in the year 2000 to 2,620,000 in the year 2050 with a majority requiring LTC (Brookmeyer et al., 2007).

Traditionally, the younger generation has always taken care of the elderly. However, several developments make this less likely. Firstly, medical progress also implies that the period of care increases beyond the capacity of the younger generation. It makes a difference whether one has to take care of the mother or father for 1–2 years or for a decade. Secondly, the demographic transition also implies that the middle generation is not as strong as it was in former times. The quotient between the elderly and the middle generation is getting higher leaving more elderly per caregiver. Thirdly, in former times many German women did not work outside their homes but took care of the children and the elderly. Today, almost all women and men work so that there is only very limited time to care for the elderly. Finally, there might also be a slight change of the value system with more emphasis on self-actualisation and less focus on serving others.

Consequently, more and more elderly require LTC from professional services. In the beginning, it might be sufficient to be visited regularly by a mobile nursing service, while up to 50% of elderly finally end-up in a nursing home. For many the costs of these homes go beyond their ability to pay. Thus, getting old induces the risk of catastrophic expenditure for LTC. The logical consequence is an LTC insurance which pools the financial risk of LTC.

The ageing population of Germany increasingly requires long-term care. The financial risk goes beyond the ability of the individual and calls for a solidarity-based pooling of risks. The underlying problem that inspired the launch of the social LTC insurance persists.

4.4.2 Development of Schemes

As stated before, some 72 million German citizens are covered by the social LTC insurance while some 9 million are under the protection of a private LTC insurance (2018) (BMG 2018c). The number of beneficiaries per year has strongly increased and is now above 3,000,000 people. Ambulatory services are still the main benefit, but also nursing homes are more and more required (see Fig. 4.3).

By the end of 2016, most ambulatory beneficiaries were on level I (64.5%) or level II (27.7%), while level III (7.8%) and hardship cases (2.1%) were still rare. For beneficiaries in nursing homes the respective figures were 43.7% (level I), 37.4% (level II), 18.9% (level III) and 5.1% (hardship cases), i.e., the level of care tends to

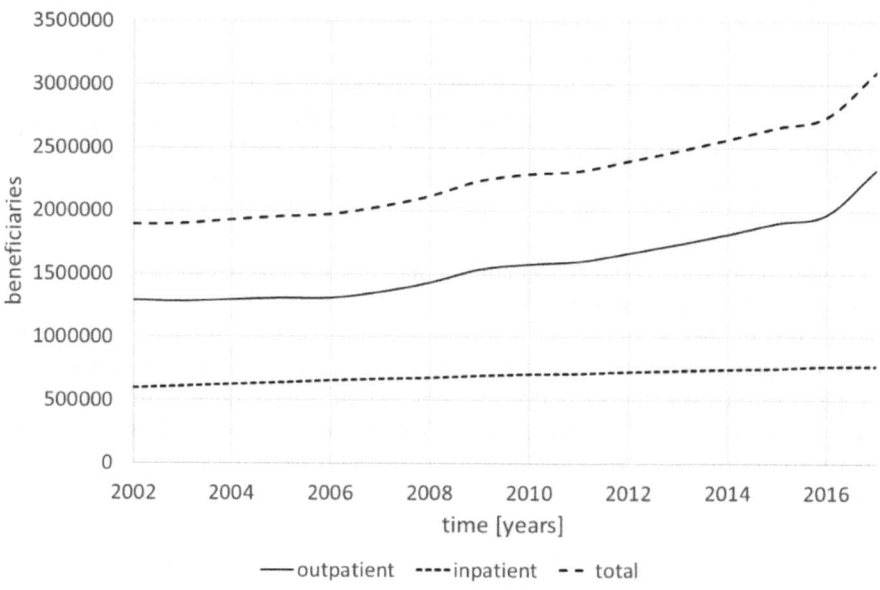

Fig. 4.3 Beneficiaries of social LTC insurance (*Source* own chart based on data from BMG 2018c)

be higher for clients requiring nursing homes than for ambulatory services. Since the introduction of care grades (1.1.2017) the picture has not changed (see Table 4.3).

As Fig. 4.4 shows, the expenditure of the social LTC insurance has increased dramatically over the last 15 years. In particular the expenditure of ambulatory care has grown and are now higher than the cost for nursing homes.

Table 4.3 Beneficiaries of LTC insurance by care grades (30th June 2017)

Care Grade	Outpatient		Inpatient		Total	
	Absolute	%	Absolute	%	Absolute	%
1	75,607	3.2	3,027	0.4	78,634	2.5
2	1,211,569	52.0	191,811	24.7	1,403,380	45.2
3	651,122	28.0	231,233	29.8	882,355	28.4
4	280,731	12.1	222,075	28.6	502,806	16.2
5	108,770	4.7	127,894	16.5	236,664	7.6
total	2,327,799	100.0	776,040	100.0	3,103,839	100.0

Source Own table based on BMG 2018c

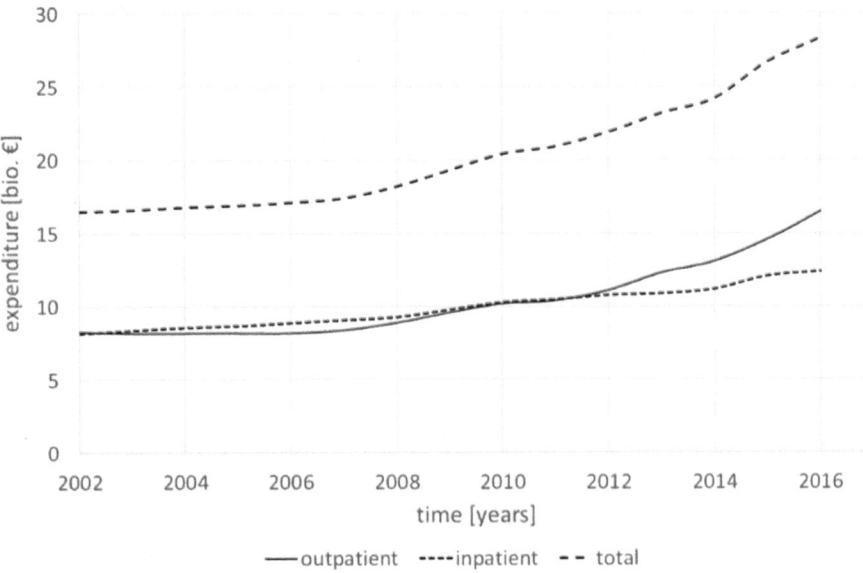

Fig. 4.4 Expenditure of social ltc insurance (*Source* Own chart based on data from BMG 2018c)

4.4.3   *Reforms*

As stated before, the social LTC insurance has undergone several reforms to accommodate major changes of the care situation. This includes:

- Dementia: Conditions caused by dementia were not included in the original system as people living with dementia might not suffer from major limitations in their activities of daily living. However, they still require care and in many cases people with that diagnosis cannot be taken care by the relatives (Dreier & Hoffmann, 2013). The new care grades give better account to the reality of an ageing population with a high number of people living with dementia.
- Support of relatives: A growing number of relatives have to give up their paid jobs to take care of their relatives. The social LTC insurance pays for their health insurance and unemployment insurance for the period of caregiving.
- Strengthen ambulatory services: Nursing homes are very expensive and frequently second choice of the elderly who want to stay in their home if possible. Thus, providing equipment and ambulatory services is strengthened more and more.

As the population of Germany is ageing, more reforms will become necessary—it remains a work in progress.

> The long-term care insurance in Germany is a great success offering social protection to the entire population. The number of beneficiaries and the cost of social long-term care insurance have continuously increased and challenge the sustainability of the system.

4.5   Discussion

As stated earlier, the German LTC insurance is a work in progress and constantly under reform. Some of the challenges cannot be solved completely but will require compromises acceptable to the population. The following challenges can be identified:

- Increasing demand: As show in Sect. 4.1, the German population is rapidly ageing, and more and more people will require long-term care. The LTC insurance addresses the symptoms of that structural problem, not the causes. Germany would need a younger population where more healthy people of the working age can take care of the elderly. However, this problem is very difficult to address, and Germany will have to face the rapid increase of demand and even over-proportional increase of costs of LTC. For the time being, the most promising instrument to sustain the existing insurance is to keep people healthy for as long as possible without requiring LTC. This would require more investments in prevention and health promotion. However, the German social LTC insurance is not permitted to finance any preventive or promotional programmes. This would be entirely in the responsibility of the social health insurance, but both types of insurances are thoroughly separated. Thus, the demand is likely to increase steadily with even more exploding costs.
- Capacity: The increasing demand will meet a limited supply of long-term care. Already today, Germany faces a severe shortage of nurses on all levels (hospitals, ambulatory services, nursing homes). 30 years ago, nursing was a profession of high social esteem, but this seems to have changed. More and more nursing schools complain that they cannot fill their training positions. Consequently, the number of nurses available will decline and the willingness to work with old, multi-morbid long-term patients is low. Even if funds were sufficiently available, the capacity might not be enough to

provide all beneficiaries with professional care. At the same time, more and more people of care-giving age-sets are working in paid jobs without a chance to take care of their ageing relatives. Germany is running in a capacity problem.

- Financing: The social LTC insurance of Germany was in a deficit for the first years, but now it can break-even. However, prospects are threatening as the demand will increase and costs are likely to explode. The willingness to increase the contribution (and in particular the employer's contribution) is limited as this has negative consequences for the German industry. Germany will likely face a deficit of its LTC insurance within a few years. At the same time the benefits are insufficient to eliminate the financial burden for the clients. The need for long-term care is still a high risk of poverty for the individual and their families.

- Administration: Finally, the social LTC insurance in Germany is very complex with different regulations and processes. As stated before, every reform has increased the complexity. The conflict between fairness (having the best solution for everybody) and simplicity (making rules understandable) is unsolved and German decision-makers tend to address it by adding regulation after regulation and complexity to complexity.

The consequence of these problems is that new reforms are to be expected. The system described in this paper will not be maintained for a long time, but amendments will be made soon. The German social LTC insurance might merely look like a financing tool—but in reality it is a constant call for a stronger public health effort: to keep the ageing population healthy and autonomous as long as possible. This is the real challenge.

References

Bärnighausen, T., & Sauerborn, R. (2002). One hundred and eighteen years of the German health insurance system: Are there any lessons for middle-and low-income countries? *Social Science & Medicine, 54*(10), 1559–1587.

Beckers, D., & Buck, M. (2014). Activities of daily living. In S. Adler, D. Beckers, & M. Buck (Eds.), *PNF in practice* (pp 293–300). Springer.

BMG, Bundesministerium für Gesundheit. (2018a). Die Leistungen der Pflegeversicherung im Überblick.

BMG, Bundesministerium für Gesundheit. (2018b). Neue Pflegegrade ab 2017. Retrieved from: https://

BMG, Bundesministerium für Gesundheit. (2018c). Zahlen und Fakten zur Pflegeversicherung.

Brookmeyer, R., Johnson, E., Ziegler-Graham, K., & Arrighi, H. M. (2007). Forecasting the global burden of Alzheimer's disease. *Alzheimer's & Dementia: THe Journal of the Alzheimer's Association, 3*(3), 186–191.

Campbell, J. C., Ikegami, N., & Gibson, M. J. (2010). Lessons from public long-term care insurance in Germany and Japan. *Health Affairs, 29*(1), 87–95.

Dreier, A., & Hoffmann, W. (2013). Dementia Care Manager für Patienten mit Demenz. *Bundesgesundheitsblatt-Gesundheitsforschung-Gesundheitsschutz, 56*(10), 1398–1409.

Eisenmenger, M., Pötzsch, O., & Sommer, B. (2006). *Bevölkerung Deutschlands bis 2050. 11. koordinierte Bevölkerungsvorausberechnung.* Wiesbaden: Statistisches Bundesamt. Retrieved from: https://www.destatis.de/DE/Themen/Gesellschaft-Umwelt/Bevoelkerung/Bevoelkerung svorausberechnung/Publikationen/Downloads-Vorausberechnung/bevoelkerung-deutschland-2050-presse-5124204069004.pdf;jsessionid=1FEB1F5019EE5CA99E6D12C01FBB5857.liv e712?__blob=publicationFile. Accessed November 03, 2019.

Engelman, L. (2008). Long-term care insurance. *Google patents.* Retrieved from: https://patent images.storage.googleapis.com/2f/d5/6f/dead5058244ae8/US20080215376A1.pdf. Accessed March 13, 2019.

Fischer, S. (2015). Die deutsche Sozialversicherung zwischen Beitrags-, Steuer- und privater Finanzierung. In L. Mülheims, K. Hummel, S. Peters-Lange, E. Toepler, & I. Schuhmann (Eds.), *Handbuch Sozialversicherungswissenschaft* (pp. 827–842). Springer. https://doi.org/10.1007/978-3-658-08840-8_52

Fleßa, S. (2018). *Systemisches Krankenhausmanagement.* DeGruyter.

Fleßa, S., & Greiner, W. (2013). *Grundlagen der Gesundheitsökonomie: Eine Einführung in das wirtschaftliche Denken im Gesundheitswesen.* Springer. https://doi.org/10.1007/978-3-642-309 19-9_1

Hinselmann, D. (2016). Einführung in die Begutachtung nach dem neuen Pflegebedürftigkeitsbegriff. *Das Gesundheitswesen, Issue, 4*(78), V16. https://doi.org/10.1055/s-0036-1578831

Johansson, P.-O. (2000). Properties of actuarially fair and pay-as-you-go health insurance schemes for the elderly. An OLG model approach. *Journal of Health Economics, 19*(4), 477–498.

Kronenberg, V. (2015). Sozial(versicherungs)politik in der Bundesrepublik Deutschland. In L. Mülheims, K. Hummel, S. Peters-Lange, E. Toepler, & I. Schuhmann (Eds.), *Handbuch Sozialversicherungswissenschaft* (pp. 55–68). Springer.

Richter, R. (2017). *Die neue soziale Pflegeversicherung-PSG I, II, I: Pflegebegriff, Vergütungen, Potenziale* (2nd ed.). Vincentz Network.

Sozialversicherung kompetent. (2018). *Beitragssatz Pflegeversicherung 2018.* Retrieved from: https://sozialversicherung-kompetent.de/pflegeversicherung/versicherungsrecht/793-bei tragssatz-pflegeversicherung-2018.html. Accesed February 28, 2018.

Stiglitz, J. E., & Rosengard, J. K. (2015). *Economics of the public sector: Fourth international student edition.* WW Norton & Company.

Zweifel, P., Breyer, F., & Kifmann, M. (2009). *Health economics.* Springer. https://doi.org/10.1007/978-3-540-68540-1_7

Steffen Fleßa is head of department of health care management at the University of Greifswald, Germany. After his training in business administration (focus on health care management) he gained practical experience in Tanzania for five years. Later he was research assistant at the department of medical economics, University of Erlangen-Nürnberg. Further steps in his career were a professorship of nursing administration (Evang. University of Applied Sciences Nürnberg), professorship of international health economics (Medical School, University of Heidelberg) and full professor of health care management at the University of Greifswald. He was dean of studies and vice dean of the faculty of law and economics. Currently he is vice rector of the University. His main research interests are in the field of quantitative modelling, health care management and non-profit management in Germany as well as in low and middle-income countries. He has research and working experience in Tanzania, Kenya, Burkina Faso, Venezuela, Vietnam, Cambodia and Indonesia. He has published more than 400 books and papers in national and international journals.

Chapter 5
Exploring the Establishment of Statutory Social Long-Term Care Insurance in China

Chun Ding and Dan Liu

5.1 The Need to Establish Statutory Social Insurance for Long-Term Care

5.1.1 The Unrelenting Advance of an Ageing Society

China has faced an ageing society since the beginning of the twenty-first century. In 2001, the country's number of over 65-year-olds reached 7.065% of the total population, over the 7% threshold to be classed as an ageing society. Since then, it has surpassed the global average level of ageing. In 2018, this ratio reached 11.9% (National Bureau of Statistics of China, 2019). In less than 20 years, the elderly population has increased by nearly 4.8 percentage points, whilst the world average[1] only rose by approximately 1.9 percentage points. The rapid rate of ageing has shocked the world. According to predictions by the United Nations (UN), by 2020, China's population of 65-year-olds will reach 167 million, accounting for about a quarter of the world's elderly population. By then, the number of oldest-old[2] in China will increase to about 29 million, and the number of elderly living alone and in an empty nest will increase to about 118 million (State Council, 2017). It is predicted that the rate of ageing will peak around 2040. In the face of an unrestrainedly ageing

---

[1] The number of over 65-year-olds worldwide made up 6.96% of the global population in 2001, rising to 8.87% in 2018 (World Bank 2019).

[2] 80-year-olds and over.

C. Ding (✉)
Institute of World Economy, Fudan University, Shanghai, China
e-mail: chunding@fudan.edu.cn

D. Liu
School of Economics, Fudan University, Shanghai, China

© The Author(s) 2025
Deutsche Gesellschaft für Internationale Zusammenarbeit (GIZ) GmbH et al., (eds.),
Sustainable Aging, https://doi.org/10.1007/978-3-662-69139-7_5

society, China must also face the social security pressures that come with it, such as elderly medical care, nursing care, and elderly care.

According to statistics from the "Fourth Survey on the Living Conditions of China's Urban and Rural Elderly" released by the National Office for the Elderly in 2015, the number of disabled elderly in China is about 40.63 million, accounting for 18.3%[3] of the elderly population. However, the structure of the elderly population in China is relatively young. At the end of 2015, the young-old sub-group (60–69 years old) accounted for 56.1% of the elderly population, the middle-old subgroup (70–79 years old) accounted for 30.0%, and the oldest-old subgroup (80 years old and over) accounted for 13.9%. This means that the degree of ageing in China will further deepen in terms of age structure and number of groups. Therefore, challenged with an increasingly large number of elderly groups, it is crucial to urgently solve the problem of care for them. The process of improving the social security system in China is also an arduous task that must be overcome.

5.1.2 Shrinking Family Structure

Rapid ageing in China is mainly due to the increase in life expectancy and the significant decrease in fertility rates. Since the reform and opening, the improvement of the economy has greatly enhanced living standards, and the life expectancy per capita has also increased from 67.77 years in 1981 to 76.34 years in 2015 (see Table 5.1). At the same time, due to the implementation of the family planning policy, China's fertility rate fell from a peak of 23.33‰ in 1987 to around 12‰, and the average family size decreased from 4.41 per household in 1982 to an average of 3.1 people per household in 2010. In China, one-child-families gradually became the norm and family units are characteristically smaller, consisting of fewer children. In 2015, the elderly in China had an average number of 3.0 children, of which the young-old had an average of 2.5 children, those aged 70–79 had an average of 3.4, and the oldest-old aged 80 and above had an average of 4.1 (Dang et al., 2018). The old-age dependency ratio in society climbed from 8.2% in 1982 to 16.8% in 2018, at the same time, the child dependency ratio dropped from 54.6% to 23.7% (National Bureau of Statistics of China, 2019). Along with the trend of the elderly having fewer children, and the growing phenomenon of adult children living separately from their parents, elderly groups living alone have also become a key concern for society, especially those who live alone and have lost the ability to care for themselves. More people are needed to take care of the disabled elderly to provide care and daily assistance. The disabled elderly are currently facing very real struggles; their quality of life has seriously declined and life is difficult in later years. Therefore, when the family's ability to provide care is weakened, society must take over the reins.

---

[3] Here, the elderly population is calculated as the population aged 60 or older.

Table 5.1 Life expectancy of China's population

Year	Aggregate	Men	Women
1981	67.77	66.28	69.27
1990	68.55	66.84	70.47
1996	70.80		
2000	71.40	69.63	73.33
2005	72.95	70.83	75.25
2010	74.83	72.38	77.37
2015	76.34	73.64	79.43

Source Own chart based on National Bureau of Statistics of China (2019)

5.1.3 Large Demands on Nursing Take Up Precious Medical Resources

As a country with a large population, China's medical resources are very scarce. Unbalanced regional development and the dualistic medical structure in urban and rural areas make it more and more difficult to "see a doctor". However, with China's current social security system, there is a lack of insurance for nursing care, and the only statutory medical insurance bank is taking matters into its own hands. According to a survey conducted by Huang and Liu in 2005, more than 70% of the elderly chose medical services as their most requested option (Huang & Liu, 2006). Faced with this situation, many elderly people who need only elderly care without medical treatment have no choice but to occupy hospital beds. Moreover, preventive care for the elderly in China is lacking; the elderly always seek medical care after falling ill, resulting in a waste of medical resources and low efficiency. Therefore, the introduction of statutory social long-term care insurance will effectively divest care needs from medical needs, not only to effectively alleviate the shortage of medical resources, but also to better solve the health care problems of the elderly.

5.2 Preliminary Steps in the Implementation of China's Statutory Social Insurance for Long-Term Care

To actively respond to the series of problems brought about by an ageing society, the Fifth Plenary Session of the 18th CPC Central Committee and the National 13th Five-Year Plan outlines the task needing to explore and establish a long-term care insurance system and conduct pilot implementation of long-term care insurance. In June 2016, the Ministry of Human Resources and Social Security issued the "Guiding Opinions on Piloting the Long-Term Nursing Insurance System" and decided to carry out long-term care insurance pilot programmes in 15 cities and a further two provinces[4] across the country, striving to basically build a policy framework for a long-term care insurance system during the "13th Five-Year Plan" period. The goal is to explore the establishment of a social insurance scheme that provides funds for social assistance and mutual aid and provides a social security system to provide guarantees for funds or services for the basic life care of long-term disabled persons and medical care needed for daily life, using a pilot period of 1–2 years to explore the path of reform and accumulate experience. Focus is on long-term care insurance coverage, insurance contributions, treatment payment and other policy systems; standards and management methods to categorise and rate nursing care needs; measures to evaluate the quality of various types of long-term care services and nursing staff service, management protocol and fee settlement; and systems for long-term care insurance service management specifications and operational mechanisms (NPC, 2017).

5.2.1 Overview of China's Long-Term Care Insurance Pilot Scheme

The first batch of 15 cities in China to participate in a long-term care insurance scheme are Chengde (Hebei province), Changchun (Jilin province), Qiqihar (Heilongjiang province), Shanghai, Nantong and Suzhou (Jiangsu province), Ningbo (Zhejiang province), Anqing (Anhui province), Shangrao (Jiangxi province), Qingdao (Shandong province), Jingmen (Hubei province), Guangzhou (Guangdong province), Chongqing, Chengdu (Sichuan province), and Shihezi (Xinjiang Production and Construction Corps). The implementation of the long-term care insurance in the first batch of the pilot cities is shown in Table 5.2.

On 10th September 2020, the National Healthcare Security Administration issued "Guiding Opinions on Expanding the Pilot programme of Long-Term Care Insurance System (Draft for Comments)" (hereinafter referred to as the "Opinions"), aimed at expanding the list of pilot cities for the development of the long-term care insurance system. A further 14 cities have been added to the pilot programme. The second batch of pilot cities (regions) are: Shijingshan District in Beijing, Tianjin City, Jincheng City in Shanxi Province, Hohhot City in Inner Mongolia Autonomous Region, Panjin City in Liaoning Province, Fuzhou City in Fujian Province, Kaifeng City in Henan Province, Xiangtan City in Hunan Province, and Nanning City in the Guangxi Zhuang Autonomous Region, Buyi and Miao Autonomous Prefecture in Southwest Guizhou Province, Kunming City in Yunnan Province, Hanzhong City in Shaanxi Province, Gannan Tibetan Autonomous Prefecture in Gansu Province, and Urumqi City in

---

[4] Though it is considered the norm for each province to have just one pilot city, Jilin province (Changchun, Jilin, Tonghua, Songyuan, Meihekou, Hunchun) and Shandong province (Jinan, Qingdao, Zibo, Zaozhuang, Dongying, Yantai, Weifang, Jining, Tai'an, Weihai, Rizhao, Linyi, Dezhou, Liaocheng, Binzhou, Heze) are key roll-out provinces for long-term care insurance. As a result, multiple prefecture-level cities in the two provinces have become pilot cities.

Table 5.2 Overview of the first batch of pilot cities adopting long-term care insurance

Pilot city	Start date	Insured party	Service content	Source of funds	Payment for treatment
Qingdao	July 2012	Urban workers, urban and rural residents	Specialised hospital care, nursing home care, home-based residential care, community care	Medical insurance + fund allocation	Proportional reimbursement; or by carrying out daily bed management[5]
Shanghai	July 2013	Urban workers, (Oldest-old sub-group)	Medical care	Medical insurance fund allocation and individual contributions	Proportional reimbursement
	January 2017	Urban workers, urban and rural residents (over 60 s)	Community residential care, retirement home care, and hospital care	Financed by the individual, the employer and the government	1. Community home-based care: The long-term care insurance fund covers 90% of costs incurred by insured parties receiving community home-based care services within the valid evaluation period 2. Institutional care: The long-term care insurance fund covers 85% of costs incurred by insured parties in institutional care that adhere to the scheme's reimbursement conditions 3. In-patient medical care: reimbursed in line with medical insurance conditions

(continued)

[5] In Qingdao, the fixed daily cost for an insured person who receives medical care at designated nursing institutions or at home is RMB 60 per bed (EUR 8.50); for an insured person who receives special medical care in a Grade II hospital the fixed daily cost is RMB 170 per bed (EUR 24); and for insured persons who receive medical care in Grade III hospitals the fixed daily cost is RMB 200 per bed (EUR 28). In 2015, Qingdao city adjusted its entire standards for elderly long-term care insurance, carrying out daily bed management to standardise costs for specialised medical care, nursing home medical care, and home medical care expenses, and conducting annual reviews of community health visit expenses. The authorities adjusted the cost of specialised medical care in six of the city's districts to RMB 170 per day (EUR 24), daily expenses for medical care in nursing homes was set at RMB 65 (EUR 9), and the daily cost of home medical care was set at RMB 50 (EUR 7). For workers insured for community health visits and Class I insurance their contributions are RMB 1600 per year (EUR 224), and for those insured under Class II contributions are RMB 800 per year (EUR 112). Standards for fund appropriation are linked to the number and quality of care services.

Table 5.2 (continued)

Pilot city	Start date	Insured party	Service content	Source of funds	Payment for treatment
Changchun	May 2015	Urban workers and urban residents	Daily care and medical care	Medical insurance and fund allocation	Proportional reimbursement is controlled at around 70%
Nantong	January 2016	Urban workers and urban residents	Daily care and medical care	Set at around 3% of last year's per capita disposable income of urban residents in Nantong City, the fundraising standard is tentatively set at RMB 100 p.p. p.a. (EUR 14), of which individual contributions are RMB 30 p.p. (EUR 4), the medical insurance pool contributes RMB 30 p.p. (EUR 4), and government subsidies cover RMB 40 p.p. (EUR 5.60)	1. Care services received in designated institutions are reimbursed at 60%, and at the same time, one can benefit from hospitalisation under basic medical insurance according to regulations 2. Elderly care services in designated institutions are reimbursed at 50% 3. For designated organisations providing at-home care services, the monthly limit is tentatively set at RMB 1200 (EUR 168)
Shihezi	January 2017	Urban workers, urban and rural residents	Daily care and medical care	Funds raised via multiple channels, including individual contributions, employer contributions, basic medical insurance pooled funds, financial subsidies and income from the expansion of the public welfare fund	Reimbursed by the care insurance fund at a rate of 70%, the monthly payment limit is RMB 750 (EUR 105)

(continued)

Table 5.2 (continued)

Pilot city	Start date	Insured party	Service content	Source of funds	Payment for treatment
Shangrao	January 2017	Firstly workers, then extending to residents	Small subsidies for residential care, care given at-home and institutional care	RMB 100 p.p. p.a. (EUR 14), of which individuals contribute RMB 40 (EUR 5.60); Medical insurance pooled funds transfer RMB 30 p.p. (EUR 4); and employers contribute RMB 30 p.p. (EUR 4)	Proportional reimbursement
Anqing	January 2017	Urban employees participating in basic medical insurance	Institutional care, home care and residential care	Individual contributions, balance transfer of basic medical insurance for urban employees pooled funds and financial subsidies	Institutional medical care: 60% reimbursement (RMB 50 ceiling, EUR 7). Institutional elderly care: 50% reimbursement (RMB 40 ceiling, EUR 5.60) At-home care insurance: RMB 750 ceiling (EUR 105) Residential care: RMB 15 /day (EUR 2) reimbursement
Chengdu	July 2017	Firstly workers, then extending to residents	Life care, nursing care, risk prevention and health maintenance	Funds raised via multiple channels, including account transfer of social Basic Medical Insurance pooled funds, and contributions by the employer, government, and charities	1. In long-term institutional care, the fixed payment standard is determined according to the disability level corresponding to 70% of the care expenses 2. For residential long-term care at home, the fixed payment standard is determined according to the disability level corresponding to 75% of the care cost[6]

(continued)

[6] After the accumulated payment period of Chengdu's Long-term Care Insurance has reached 15 years, the payment standard will increase by 1% for every 2 years. The proportion of long-term care insurance fund payments cannot exceed 100%. Before the start of the long-term care insurance scheme, the actual payment period for participating in basic medical insurance was regarded as the long-term care insurance contribution period.

Table 5.2 (continued)

Pilot city	Start date	Insured party	Service content	Source of funds	Payment for treatment
Jingmen	January 2017	Those participating in basic medical insurance	Residential care, retirement home care and medical care	Individual contributions, transfer of social Basic Medical Insurance pooled funds, and financial subsidies. In 2016 the level of funds was set at RMB 80 p.p. (EUR 11), with RMB 30 (EUR 4) coming from individual contributions, RMB 20 (EUR 2.80) from funds and RMB 30 (EUR 4) from government financing[7]	Non-full-time residential care: limited to 40 RMB p.p. (EUR 5.60) and the fund is paid in full. Full-time residential care: limit of RMB 100 (EUR 14), whereby the fund pays 80%. Institutional care: daily limit of RMB 100 (EUR 14), whereby the fund covers 75%
Chengde	July 2017	Insured persons participating in medical insurance for urban employees throughout the city	Nursing care services at designated organisations	The annual funding standard is tentatively set at 0.4% of the previous year's salary for the insured (including retirees), 0.2% borne by the basic medical insurance fund for urban employees, 0.15% in individual contributions by the insured (including retirees), and 0.05% by government financial assistance	The payment standard is set per bed allocation: Specialised care at medical institutions: RMB 60 per bed/day (EUR 8.50). Services at designated care institutions: RMB 50 (EUR 7) per bed/day. At-home care given by designated care institutions: RMB 40 per day (EUR 5.60). If the actual calculation of costs of services provided by designated care institutions is lower than the standard, 70% of the cost of care that meets the scope of reimbursement for long-term care insurance is paid

(continued)

[7] Jingmen City has established an incentive mechanism for participating in insurance contributions, by linking the period of payment to the level of treatment, encouraging early participation in the scheme and continuous contribution. The specific measures are as follows: for cumulative payments of 15 years or more, the level of treatment increases by 4%; for cumulative payments of more than 30 years, the level of treatment increases by 6%; for cumulative contributions of more than 45 years, the level of treatment increases by 8%; and for accumulated contributions of more than 60 years, the treatment level will increase by 10%.

Table 5.2 (continued)

Pilot city	Start date	Insured party	Service content	Source of funds	Payment for treatment
Guangzhou	August 2017	Firstly workers, then extending to residents, and finally to full coverage	Institutional care and residential care	RMB 130 p.p. p.a. (EUR 18) according to standards of financing and corresponding treatments. The medical insurance fund is pooled by measuring the long-term care insurance income and expenditure needs of the following year	Proportional reimbursement: Institutional care is reimbursed at 75%, and residential care at 90%
Suzhou	October 2017	Urban workers, urban and rural residents	Medical institution inpatient care, retirement home care and community residential care	The long-term care insurance fund raises funds in accordance with the principle of expenditure based on income, balance of payments, and a balance with a small surplus. The long-term care insurance fund consists of individual contributions, government subsidies, basic medical insurance for employees, and basic medical insurance for urban and rural residents pooled funds[8]	Fixed-rate payment: the standard for severely disabled persons is RMB 26 per day (EUR 3.60), the standard for moderate disability is RMB 20 per day (EUR 2.80); the standard for severely disabled persons in home-based residential care is RMB 30 per day (EUR 4), and for moderate disability is RMB 25 per day (EUR 3.50)

(continued)

[8] During the first phase of the pilot project in Suzhou (2017–2019): individual contribution is partly exempted from collection, the government subsidises RMB 50 p.p. p.a. (EUR7), the balance of basic medical insurance for employees is transferred at 70 RMB p.p. p.a. (EUR 10), and the balance of basic medical insurance for urban and rural residents is transferred at RMB 35 p.p. p.a. (EUR 5). For the second phase (2020 and beyond), the medical insurance fund that covers persons suffering from long-term disability and medical care that is closely linked to life care is integrated into the long-term care insurance system.

Table 5.2 (continued)

Pilot city	Start date	Insured party	Service content	Source of funds	Payment for treatment
Qiqihar	October 2017	City-level participation (excluding the Meilisi Daur ethnic minority area) those participating in basic medical insurance for urban workers (including flexible employment)	Medical care, elderly care and residential care	The funding standard is RMB 60 p.p. p.a. (EUR 8.50), of which the individual insured and the medical insurance pooling fund each contribute RMB 30 (EUR 4)	According to daily quota management, the expenses within the fixed amount shall be proportionally reimbursed by the long-term care insurance fund. The daily quota for the insured receiving medical or nursing care from medical institutions or elderly care agencies whilst in medical care institutions, elderly care institutions, or in home-based care is RMB 30 (EUR 4), RMB 25 (EUR 3.5), RMB 20 (EUR 2.80), respectively. The insurance fund covers 60%, 55%, and 50% respectively
Ningbo	December 2017	(Haishu District, Jiangbei District and Yinzhou District) Insured workers participating in basic medical insurance for employees	Specialised institutional care (special care) and elderly care (hospital care)	During the pilot period, RMB 20 million (EUR 2,8 million) of funds will be raised from the accumulated balance of the urban employees' basic medical insurance pooling fund as the pilot fund for the long-term care insurance, and will be included in the long-term care insurance fund. Individuals and employers do not pay any fees for the time being	The daily quota is tentatively set at RMB 40 (EUR 5.60)

(continued)

Table 5.2 (continued)

Pilot city	Start date	Insured party	Service content	Source of funds	Payment for treatment
Chongqing	December 2017	Firstly workers, then extending to residents	Daily care and medical care	The 2018 fundraising standard is set at RMB 150 p.p. p.a. (EUR 21). Of which, the medical insurance fund subsidises (RMB 60 p.p. p.a. (EUR 8.50), and the insured worker of medical insurance for employees contributes RMB 90 p.p. p.a. (EUR 12.50)	Fixed-rate payment: The long-term insurance fund is set at a rate of RMB 50 p.p/day (EUR 7)

Source The table has been compiled based on public information. For details, please refer to government policy documents such as the pilot opinions of the long-term care insurance system in each pilot city

Xinjiang Uygur Autonomous Region. In addition to the pilot city clusters in the key provinces of Shandong and Jilin, the Chinese authorities have already reached 49 pilot cities (regions), accounting for 14.7% of the total number of prefecture-level cities in China. At the provincial level, except for Hainan, Qinghai, and the Tibet Autonomous Region, 28 provinces, autonomous regions, and municipalities in mainland China have already launched pilot programmes for long-term care insurance, representing a coverage of 90.3%.

Although it has been less than half a year since the official release of the "Opinions", the second batch of pilot cities has actively carried out pilot testing, exploring the feasibility of long-term care insurance rules in line with local economic and social conditions, by conducting research, seminars, symposia, and soliciting public comments. Some pilot cities from this second batch have even issued long-term care insurance rules (see Table 5.3).

5.2.2   Implementation Results in Pilot Cities

To date, the implementation of the long-term care insurance scheme has been smooth in all 15 pilot cities from the first batch and across the two key roll-out provinces, and the results are beginning to show. The main outcomes are summarised below:

(1)   The economic and transactional burden on the disabled elderly and their families has been reduced, with overall reimbursement rates at around 70–75% for long-term care costs adhering to the reimbursement conditions. Up until May 2020, the number of insured has surpassed 88.58 million, benefitting over 426 thousand people that same year, the rate covered by the funds has reached over 70%, an average expense of RMB 9200 per person (EUR 1290), showing early signs of success for the insurance scheme.[9]

(2)   Among the first batch of pilot cities, 10 cities[10] adopted international disability assessment standards, and 5 cities[11] developed their own disability assessment standards. The disability assessment pass rate is about 80%, combining existing international experience as well as exploring and putting into practice disability rating standards that conform to China's national conditions.

(3)   The pilot areas have also successfully acted as platforms to allocate institutional resources and purchase services to support the development of the elderly healthcare industry and the construction of the care service system, promoting supply side labour reform. According to incomplete statistics, more than 40,000 people have been employed as a direct result of the pilot scheme, and more than RMB 7 billion (EUR 981 million) of social capital has been injected (MoHRSS, 2018). A total of 3242 service agencies have been set up in the pilot cities, and

---

[9] Data source: China Medical Insurance.

[10] Chengde, Ningbo, Anqing, Jingmen, Nantong, Qiqihar, Shihezi, Guangzhou, Changchun and Chongqing.

[11] Qingdao, Shanghai, Suzhou, Shangrao and Chengdu.

Table 5.3 The development of long-term care insurance in some pilot cities from the second batch

Pilot city	Start date	Insured party	Service content	Source of funds	Payment for treatment
Beijing (Shijingshan)	15th March 2018	In-service employees, retirees and urban and rural residents covered by basic medical insurance for urban and rural employees (not including students and children)	Activities of daily living (ADL) for disabled persons and medical care services closely related to ADL are guaranteed, whereas corresponding nursing services are provided in kind	State, employer and individual contributions, with a funding ratio of 4:4:2, social donations also accepted Standard: RMB 160 p.p. p.a. (EUR 22.50)	Institutional nursing services: RMB 70/day (EUR 10), the insurance fund covers 70%, and the remaining 30% is paid by the individual Community nursing services at home: RMB 85/hour (EUR 12), the insurance fund covers 76%, and the individual 24%, limited to 30 h per month Relatives (home care aids) at-home care services are RMB 50/hour (EUR 7), the insurance fund covers 64%, the individual pays 36%, and the monthly payment is limited to 30 h
Urumqi	25th March 2019	Participants in the basic medical insurance scheme for urban employees, and gradually expanding to those covered by basic medical insurance for urban and rural residents	24-h home care 24-h care services in designated skilled nursing facility Home care visits from designated nursing agency	State, medical insurance, and individual contributions at a ratio of 2:5:3, social donations also accepted. RMB 100 p.p. p.a. (EUR 14)	24-h at-home care: 75% reimbursement calculated on a base rate (RMB 1862/month, EUR 260) 24-h designated nursing facilities reimbursed at 70% of the base rate (RMB 1737/month, EUR 244) Home care visits from designated nursing agencies, RMB 40/hour (EUR 5.6), no more than 2 h per day, 50% covered by the insurance fund and 50% by the individual (If the number of days receiving such services falls short of one month, compensation will be based on the actual number of days receiving care. 24-h at-home care is set at RMB 61/day (EUR 8.50), and 24-h institutional care is set at RMB 57/day, EUR 8)

(continued)

Table 5.3 (continued)

Pilot city	Start date	Insured party	Service content	Source of funds	Payment for treatment
Kunming	25th September 2020	Participants in the basic medical insurance scheme for urban employees, and gradually expanding to those insured by basic medical insurance for urban and rural residents	Integrated medical and healthcare agency services, nursing home care, at-home care	Funds raised by multiple channels including a combination of individual and employer contributions, medical insurance fund allocation, financial subsidies, social donations, and other funding methods, in accordance with the principle of "multi-party financing, sharing contribution responsibility, and unification between rights and obligations"[12]	The basis for calculation and payment of benefits is 70% of the previous year's average monthly salary in Yunnan Province. In principle, the monthly payment limit shall not exceed 70% of the benefit base rate. The overall level paid by the insurance fund is controlled at about 70% for nursing fees that meet the regulations

(continued)

[12] At the beginning of the scheme, a certain percentage was allocated from the balance of the urban employee basic medical insurance pooled funds accumulated over the years. Employer and individual contributions for in-service employees were based on 0.2% of the basic medical insurance payment base, and transferred from the urban employees' basic medical insurance pooled funds and the individual account funds; Flexible workers who participate in the basic medical insurance scheme by means of "combining pooled funds and individual accounts" shall be transferred from the medical insurance pooling fund and personal account fund at 0.2% of their payment base; flexible workers who participate in the basic medical insurance scheme by means of "individually-built pooled funds" shall be transferred from the basic medical insurance fund at 0.4% of their payments base. Since Kunming's long-term care insurance scheme was launched, a proportionate reduction of 0.2% has been applied to its urban employee basic medical insurance pooled funds and individual account funds, to ensure that the implementation of long-term care insurance does not increase the payment burden for employers and individuals. Retirees benefit from a combination of individual contributions and financial subsidies. The portion of individual payment is based on 0.2% of the retiree's medical insurance individual account transfer base and transferred from the fund that should be allocated to the individual account of the insured; the portion of financial subsidy is based on 0.2% of the retirees' medical insurance personal account transfer base. Struggling enterprises can, upon consideration by the relevant authorities, replace their retired employees' individual contributions with full financial subsidies to cover long-term care insurance.

Table 5.3 (continued)

Pilot city	Start date	Insured party	Service content	Source of funds	Payment for treatment
Nanning	27th September 2020	Employers and participants in the basic medical insurance scheme for urban employees	At-home care, home care visits and care received at designated skilled nursing facilities	In the year the pilot was launched, a lump-sum transfer of RMB 300 million (EUR 42 million) was used as the start-up fund, taken from the accumulated balance of the employee basic medical insurance pooled funds over the years, for the long-term care insurance pilot scheme The long-term care insurance payment ratio is set at 0.26%, and employer and individual contributions are at equal rates. Long-term care insurance premiums and employee basic medical insurance premiums are levied simultaneously	Severely disabled persons: RMB 2463 p.p. (EUR 346) of which: RMB 1930 (EUR 271) is the monthly quota for basic nursing benefits, and RMB 533 (EUR 75) is the monthly quota standard for professional nursing benefits (1) Daily care Designated nursing facilities or at-home nursing services are paid at 75% of the fixed standard, with the payment standard set at RMB 1447.5/month (EUR 203)

Source Compiled from publicly available information. For details, please refer to government policy documents such as the opinions on the long-term care insurance system of each pilot city the construction of long-term insurance infrastructure is showing early signs of success.

(4) Valuable experience has been accumulated via the pilot programme for promoting long-term care insurance in China. At a time when national standards have not yet been set, the pilot cities have already set reimbursement rates or fixed-rate payment standards, which can serve as a good reference for subsequent national assessments. At the same time, with the advancement of social care insurance, the evaluation criteria for disability levels have also been developed. The operation mechanism of specific care programmes and long-term insurance, including the training of corresponding operating agencies, are the successes achieved during the implementation of the pilot project.

5.2.3 Current Problems and Breakthroughs of Long-Term Care Insurance in Pilot Cities

5.2.3.1 Nursing Care Problems in Rural Areas Remain Unresolved

After the reform and opening, China has produced a large floating population where the main tendency has been for surplus labour from rural areas to move to urban areas. This has resulted in a drain of the population in rural areas, adding to the severity of the situation for the "empty nest" elderly. Furthermore, due to an imbalance in regional development, China's rural areas are lagging cities in terms of infrastructure, medical services, and care institutions, with human resources also in short supply. New rural cooperative medical insurance has not fully integrated urban and rural medical care with urban medical insurance in all regions across the country, therefore, many rural populations are excluded from the insured target. Thus, the promotion of care insurance in rural areas, as well as the construction of care facilities and training of relevant personnel will become the next big challenge to overcome for long-term care insurance in China.

5.2.3.2 Application Procedures for Treatment Are Cumbersome and Evaluation Schemes Are Not Thorough

When looking at the implementation rules of long-term care insurance in pilot cities, the main review method used for the application of care insurance in China is mostly based on rating, scoring or disability levels, and some even attach conditions for application of treatment of no less than 6 months or 180 days. Some even require evaluations and submissions through layers of institutions before finally being able to obtain the qualification for long-term care insurance. Although such bureaucratic approval and strict application conditions greatly reduce the moral hazard of the insured, the cumbersome procedures may also discourage people from participating in the scheme, affecting the promotion of care insurance and its operational efficiency.

Therefore, establishing a unified and reasonable evaluation system and an efficient auditing organisation as soon as possible is also an important prerequisite for the future implementation of long-term care insurance nationwide.

5.2.3.3 Care Insurance is Still Closely Linked to Medical Insurance, and Problems Exist in the Funding Ratio

According to the current situation in pilot cities, many of the financing channels for long-term care insurance in China come from the allocation of existing medical insurance funds, thus, long-term care insurance relies on the original medical insurance system both in terms of funding and insured groups. At the same time, financial subsidies also account for a large proportion, while individual and employer contributions are obviously insufficient.

Considering experience in foreign countries, on the one hand, care insurance needs to be completely separated from medical insurance so as to fully exert its role and to reduce the burden of medical insurance payments. The Netherlands, Germany, Japan, and Korea all promulgated laws and regulations related to long-term care insurance in 1968, 1994, 2000 and 2008 respectively, and established an independent long-term care insurance system to remove elderly health care, rehabilitation, and nursing services from the medical insurance system (Li & Zhang, 2018). On the other hand, looking at the financing structure of care insurance in developed countries, the proportion of government financing has witnessed a gradual reduction and shift towards a fixed contribution. However, China is ageing rapidly. It is estimated that by 2050, the number of elderly people requiring long-term care will reach 107.47 million. As the ageing process intensifies, the amount of expenditure will increase rapidly, and if the current financing structure is maintained, it will undoubtedly increase the expenditure burden and financial pressure on the medical insurance fund. Since the current premium rates for the five categories of social insurance for employees in China have reached 39.25%, ranking 13th (Zhao, 2016) among 173 countries and regions, there is a rush to collect premiums for the newly established long-term care insurance, which may be counterproductive. However, to ensure the sustainable development of long-term care insurance, there needs to be financing made up of reasonable individual and employer contributions, which will naturally lead to problems of how to coordinate with the existing rates of the five categories of social insurance premiums. Equally there needs to be continued exploration of reasonable fundraising structures for the care insurance fund, for which a solution is required in the next round of social security reforms.

5.2.3.4 The Level of Care Provision Needs to be Urgently Improved

The establishment of long-term care insurance is still in its infancy in China. To sustain its development, one needs not only the skeleton, to improve the institutional framework, but also the flesh and blood—the process of care provision. Insufficient levels of nursing are a major bottleneck restricting the development of China's elderly care service. Yaozhen Zhu, Deputy Director of the National Office for Ageing, said that there are a small number of people engaged in professional long-term care services—they tend to be older in age, their level of education is low, and their business skills are lacking. Less than 10% of elderly care staff are certified to work. According to international practice, one member of the nursing staff is required for every three elderly people, and based on the existing 40.6 million disabled elderly people in China, the current demand for nursing staff is at least 10 million. According to data up to the end of 2017, there were only 300,000 people in China who had obtained the qualification certificate to be an elderly care worker, which is extremely far from the number being able to meet market demands. The dilemma of "having insurance, but having no staff" casts a shadow over the sustainable development of future care insurance.

5.3 Key Development Trends in China's Long-Term Care Insurance

5.3.1 Accelerate the Legislative Process of Long-Term Care Insurance

In the process of exploring the establishment of a long-term care insurance system, China's law-making process is clearly lagging that of other countries such as Japan and Germany where laws have preceded a care insurance system. China's Social Insurance Law implemented in 2011 does not explicitly incorporate care insurance into the social security policy framework, but accelerating the legislative process of long-term care insurance and enabling long-term care insurance to enter the social security legal system will provide a unified path for the development of long-term care insurance, and accelerate the development process in China.

5.3.2 Expand the Scope of Insurance Coverage and Improve the Level of Protection

With the accumulation of experience in pilot cities, the scope of participation will gradually expand, regional coverage will gradually be extended to the whole country, and more groups will be covered until all receive coverage. The scope of protection will be long-term disabled people, not just benefitting the elderly, but also including young people suffering from disability after an accident, and children with disabilities caused by congenital diseases, gradually improving the level of protection.

5.3.3 Commercial Insurance Becomes an Important Supplement to Long-Term Care Insurance

Commercial insurance will also follow the national expansion of statutory long-term care insurance ushering in the development of care insurance, to meet the different levels of care needs of different groups of people and become an important supplement to long-term care insurance.

5.3.4 Coordinated Development of Residential Care and Institutional Care

Due to the influence of traditional family values in China, home-based residential care is still the preferred option. At the same time, the service model provided by care institutions will proceed from various practical situations, according to the different needs of the care recipients, towards a diversified development of home care, institutional care, and health visits. It will drive the effective integration of long-term care insurance and medical care.

5.3.5 Strengthen the Nursing Care Team (Ma, 2017)

Departments for health, education, civil affairs and others work together to actively develop nursing education. For nursing care practitioners, various types of education and non-degree further education programmes must be vigorously introduced, to cultivate a professional team of quality nursing practitioners.

5.3.6 Introduce Elderly Care Leave

The parents of the first generation of only children have gradually entered old age. The one-child family has caused a lot of difficulties in terms of providing care to the elderly because of lack of family members. With the rapid ageing of the population, this issue will become even more noticeable. The establishment of elderly care leave can encourage children to care for their elderly parents, and the elderly can have more time to reunite with their children, which is beneficial for the mental health and emotional needs of the elderly.

References

Dang, J. W., Wei, Y. Y., & Liu, N. N. (Eds.). (2018). *Survey report on the living conditions of China's urban and rural older persons*. Social Sciences Academic Press.

Huang, Y. H., & Liu, H. Y. (2006). Positive research on demand of senior citisen's services. *Journal of Beihua University (Social Science Edition), 2*, 91–95.

Li, C. Y., & Zhang, H. P. (2018). Long-term insurance financing models in developed countries: Comparison and experience. *Truth Seeking Journal, 3*, 69–78.

Ma, X. D. (2017). The path to reform of long-term care insurance. China insurance news (08.11.2017). Retrieved from December 5, 2020: http://insurance.jrj.com.cn/2017/11/081349 23356845.shtml

MoHRSS, Medical Insurance Department of the Ministry of Human Resources and Social Security. (2018). The pilot long-term care insurance programme is progressing well (02.05.2018). Retrieved from December 5, 2020: http://www.mohrss.gov.cn/yiliaobxs/YILIAOBXSgongzu odongtai/201805/t20180502_293342.html

National Bureau of Statistics of China. (2019). *China population and employment statistics yearbook 2019*. China Statistics Press.

NPC, The National People's Congress of the People's Republic of China. (2017). Reply of the ministry of human resources and social security to the recommendation No. 8480 of the Fifth Session of the 12th National People's Congress. Retrieved from December 5, 2020: http://www. mohrss.gov.cn/gkml/zhgl/jytabl/jydf/201711/t20171120_281955.html?keywords=8480

State Council, State Council of the People's Republic of China. (2017). Thirteenth Five-Year National Plan for Developing Undertakings for the Elderly and Establishing the Elderly Care System, Chapter 1, Section 2.

The World Bank. (2019). Population ages 65 and above. Retrieved from May 17, 2020: https://data. worldbank.org/indicator/SP.POP.65UP.TO.ZS

Zhao, P. (2016). The urgent need to curtail the five Social insurances and one housing fund and reduce labor costs. Beijing Times (30.08.2016). Retrieved from May 5, 2021: https://china.hua nqiu.com/article/9CaKrnJXkUP

Chun Ding Prof. Dr. in Economics, Jean Monnet Chair, director of Centre for European Studies, Dutch Studies of Fudan University. He has been engaged in the research and teaching of European economic integration, European Economy, mode of European welfare state and the social-economic affairs of its Member States since 1980s. Prof Ding has published nearly 150 articles and several books such as "An Empirical Comparison on Main Models of Health Care Systems in the World". He is vice president of China Association for European Studies, Chinese Society for EU Studies, board director of Chinese Association of Social Security (CAoSS), of World Economy; Schumann Chair of CES, Luxemburg University, senior re-searcher at the Institute of European integration studies (ZEI) at Bonn University; AB Member of Centre of European studies, Gothenburg University, and Sichuan University etc.; Member of global Agenda Council on Europe of World Economic Forum.

Dan Liu Ph.D. Student at the School of Economics at Fudan University with a research focus on EU economics. She studied German translation at Sichuan International Studies University and received a master's degree. Her focus areas are business, financial and news translation. She completed her bachelor's degree at Chongqing University for German Language and Literature.

Chapter 6
Qingdao Long-Term Care Social Insurance Pilot

Bei Lu and Guanggang Feng

6.1 Status Quo of the Long-Term Care System in China

Low fertility and increased longevity are primary contributors to an ageing population, leading to a shrinking labour force, a rise in dependent elders, and placing pressure on public finance. Policy makers have to be fully aware of the financial constraints associated with providing efficient and adequate support to elders in an ageing society. One big challenge of an ageing population is the health integrated "long-term care" (LTC). Although long-term care in China currently relies heavily on family support, and the LTC policy still in the early stages, it has been envisaged to have a significant increase in demand in the near future.

Literature focusing on China's LTC systems are mainly on the demand side analysis (Gu & Zeng 2006; Gu et al., 2009; Hu et al., 2015; Ma et al., 2012; World Bank, 2016; Zhu & Jia, 2009). Some have studied a LTC insurance system (Deng & Guo, 2015; Du & Wang, 2016; Jin, 2006; Lin, 2015; Wei & He, 2012). Few researches have analysed the recent development of China's LTC policy practice (Cheng & Shen, 2017; Lu et al., 2017a; Yang et al., 2016). Yang et al. (2016). Among these policy analyses, Shanghai's social health insurance model, Qingdao's LTC Insurance and means-tested Nanjing model are probably the most desirable policy options.

The Chinese central government has been giving empowerment to local jurisdictions to design LTC pilot programmes. The Ministry (and Bureaus in local governments) of Civil Affairs has been traditionally responsible for supporting the frail individuals, and they use means tested mechanism to support these needy elders (the

B. Lu (✉)
Centre for Excellence in Population Ageing Research (CEPAR), University of New South Wales (UNSW), Sydney, Australia
e-mail: lubei@unsw.edu.au

G. Feng
Anhui University of Finance and Economics, Anhui, China

© The Author(s) 2025
Deutsche Gesellschaft für Internationale Zusammenarbeit (GIZ) GmbH et al., (eds.),
Sustainable Aging, https://doi.org/10.1007/978-3-662-69139-7_6

so-called Wubaohu system).[1] The Ministry of Human Resources and Social Security has recently promoted LTC programs in the form of a social insurance to broaden coverage to all residents. While most developed countries have implemented their LTC policies, the models vary significantly. For instance, in Belgium, part of their LTC remains within the health care framework (Willemé, 2010),[2] while in Germany, the LTC insurance was primarily launched to address social needs of disabled elders.[3] In China the local LTC pilot programs take two different approaches: in Nantong the focus is on financially supporting families (through cash and government purchased packages) to access better social service for disabled elders; whereas Qingdao has initiated an LTC plan heavily focused on extending health services to residential settings. Initially, the Qingdao LTC plan had very limited social care component, primarily focused on hospital special care. The majority public social support was provided by Bureau of Civil Affairs with means test criteria.

There were sound reasons behind the LTC initiative in Qingdao. Despite China's rapid ageing, the population structure still lent itself well to informal care arrangements when the policy began. In 2010, more than 88% of the elder population had two or more children to care for them (Lu et al., 2017b, p 92) and only about 14% of them with instrumental activity of daily livings (IADLs) or activity of daily livings (ADLs) lacked family support (Lu et al., 2015, p 34). Family-provided informal care plays a crucial role in social care, alleviating demand for public social care. In China the health care system exhibits an inverted pyramid structure, where major hospitals being frequently utilized for all health care services, while primary care, minor hospital clinics or other health institutions receive fewer visits. Despite government efforts to encourage visits to the primary care centres, the desired outcome has not been achieved. As disabled elders often occupy long-term hospital beds and hospital clinic positions, a separate LTC system from the medical care structure becomes necessary to redistribute medical resource effectively. This also implies that the LTC should not merely add to overall health care expenditure but it should off-set some medical costs from traditional health resources to non-acute institutions.

It was under this circumstance the Qingdao "long term care" initiative was established, with two reforms. First, to re-distribute medical resource from the major hospital and to ensure separating the LTC from acute care. This is achieved by setting up special care types of beds for LTC recipient in the hospitals (at much cheaper price); and contracting with community and private clinics to render basic medical care services for LTC recipients at their respective residences. Both are subsidised by new arrangement from the LTC program. Second, to nurture the development of community and private clinics institutions with an expected client demand from LTC arrangements. This will enable building an effective primary/non-acute care

---

[1] Wubaohu means government provides basic food, clothes, housing, medical care and funeral to the people who have no family, no ability to work and no income.

[2] Belgium operates LTC partly under the universal health system, funded both by social security contribution and general tax.

[3] The definition of "disabled elders" varies regionally. Every pilot city has their own eligibility criteria for long-term care services. Most of them are based on Barthel's ADLs indexation, MMSE for cognitive function and medical reports.

network that will serve, not only to disabled elders, but also for the general health care needs of all residents. Though some big cities, like Shanghai, have achieved health service extensions to residential places within the medical care system, most regions in China still find it challenging to procure. Qingdao has set up an example of how to initiate the primary health care network through their LTC plan.

The Qingdao LTC pilot has been evolving and rapidly developing; after the initial purpose of medical resources redistribution was achieved, the Qingdao heath care services proceeded to recently integrate social care into their LTC Insurance (LTCI) program. In the next section, we briefly introduce the history of Qingdao LTCI.

6.2   Introduction of the Qingdao LTCI

Qingdao initiated LTCI practice in 2006, following 12 years of various reforms and regulations. Recently Qingdao issued its 2018 LTC regulation package (Qingdao, 2018) which comprised a comprehensive health and social care scheme covering primary medical care, health management, long-term nursing care, social care, palliative care, rehabilitation, and more. The focus of Qingdao LTC insurance is on supporting disabled elders, including those with dementia and Alzheimer disease, due to ageing-related issues, diseases and injuries.

6.2.1   History of LTCI Development in Qingdao

In anticipation of a rapid ageing population, the Chinese government began encouraging the development of the long-term care industry in the early 2000s by establishing goals for the number of LTC beds per 1000 elders. However, most of the so-called LTC beds are actually for self-sufficient elders.[4] Only a limited number of these beds had the capacity to accommodate elders with high care needs. Since 2006, Qingdao had piloted to support disabled elders after acute treatment under long-term care arrangements. Policy trials had encouraged community support and local clinic care.

The LTCI in Qingdao was officially launched on July 1st, 2012 through the publication of Qingdao Document No. 91, by the Qingdao Human Resources and Social Security Bureau. The policy defined three types of services: hospital high care (twenty-four hours a day) in either grade two or grade three hospitals; care directed to residential institutions, and individual homes. These services are provided at subsidised unit prices determined by the LTCI. In 2012, the daily subsidy for eligible elders nominated for nursing homes or home care was RMB 60 per day per

---

[4] For example, Qingdao had about 40,000 nursing home beds, but only about 3000 workers (Qingdao Daily, 2016).

person (EUR 8.50), while for those in second- and third-grade hospitals, the subsidy was RMB 170 (EUR 24) and RMB 200 (EUR 28) per bed per day respectively.

The 2015 policy reform further expanded the service provision. In addition to the hospital, nursing home, and home care services, which primarily serve urban citizens, a fourth pillar, "mobile clinic care" was added for rural-disabled elderly people with medical needs.

By 2016 and 2018, in new policy announcements, both dementia and Alzheimer recipient were covered by the LTCI. Both service provisions and coverage continue to improve.

6.2.2 Source of Funds and Fund Management

LTCI in Qingdao is a city-level insurance system with no individual accounts.[5] It has two membership account books: the Urban Employment Medical Insurance members, and the Rural and Urban Resident Basic Medical Insurance members. Before 2014, the sources of LTCMI fund were from medical insurance funds and fiscal transfers which partly from welfare lottery revenues.

At the end of 2014, integrated LTCI fund was established covering all citizens. The city government decided that about 20% of the accumulated Qingdao Basic Employee Basic Medical Insurance (EMI) fund balance, which was approximately RMB 1.98 billion (EUR 277 million), be transferred to the LTCI account. Every year, all Urban Employee Basic Medical Insurance members, comprising about 3.85 million people, transfer 0.5% of their individual account premium, i.e., a quarter of their 2% individual account contributions, into the LTCI account. This annual transfer totalled about RMB 500 million in 2015 (EUR 70 million). Rural and urban residents' medical insurance members (RMI) comprised about 4.92 million people, were requested to pay 10% of their total medical contribution into the LTCMI account, which totaled about RMB 300 million (EUR 42 million).[6]

In 2015, through tender, the LTCI introduced two insurance companies to manage the funds. The PICC Health Insurance Company Limited Qingdao branch managed the fund for the Employee Medical Insurance account (EMIA), and China Life Qingdao managed the funds, supporting rural and urban residents. The evaluation for service eligibility had been transferred to the insurance companies as independent certified institutions. The coordination process involved the various organisations (Lu et al., 2017b).

A special LTCI fund pool is also created in 2018, consisting of 5% of both EMI and RMI long-term care insurance account amount (see Fig. 6.1). This fund is used

---

[5] Unlike the usual social pooling medical insurance policy, China's Basic Medical Insurance consists of two parts: the social pooling part and an individual account.

[6] Unlike the usual social pooling medical insurance policy, China's Basic Medical Insurance consists of two parts: the social pooling part and an individual account.

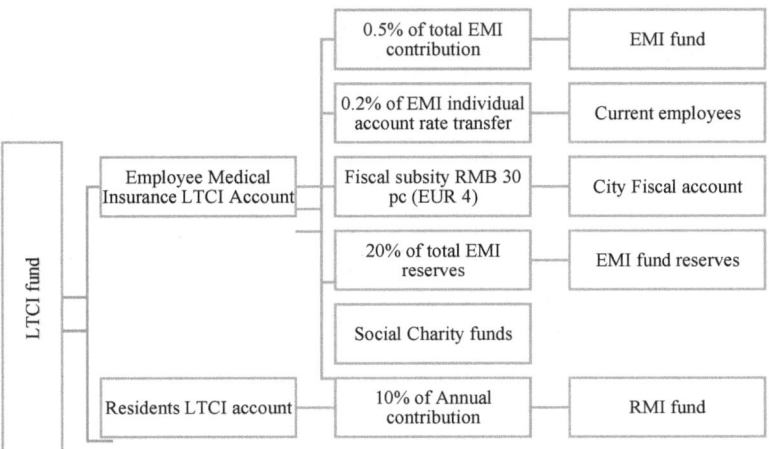

Fig. 6.1 Qingdao LTCI accounts and funding structure established in 2018. *Source* Own draft based on data from Zhejiang University (2019)

to support the training and intervention programmes for mild and minor dementias as well as cognitive decline risk cohorts.

6.2.3   Current Service Provision and Eligibility in Qingdao LTCI

The current Qingdao LTCI consists of two main parts: social care and integrated health care. Social care is defined by hours of service (in the form of hours/week to different levels of disability assessment groups) and covers about 60 items consisting of 25 basic nursing items, 17 daily living support items such as bathing, 15 rehabilitation training like swallowing and 3 medication management guidance. The integration health care includes major services following acute care, they including health management (like chronic disease management), long-term nursing care and palliative care. The LTCI also provides funds to support cognitive decline prevention programmes, aimed at helping minor and mild dementia and risk groups through functioning treatment and exercises.

Eligibility is determined by assessment criteria set up using multiple forms and examined by professional teams, which include activities of daily living (ADLs, using BARTHEL index), disease conditions, cognitive assessment (MMSE index) and others.

Table 6.1 describes the coverage and provisions. The current system provides different services and subsidies to two cohorts with different LTC needs: by ADLs and by cognitive functions (dementia).

Table 6.1 Service types to employee basic medical insurance members (EMIM), residents high-contribution medical care members (RHMIM) and residents low-contribution medical care members (RLMIM)[7]

Health status	Service types	EMIM	RHMIM	RLMIM
ADLs disabled	Hospital care	✓	✓	–
	Institutional care	✓	✓	–
	Home care	✓	–	–
	Mobile care	✓	✓	✓
Severely demented	Long-term care	✓	✓	–
	Day care	✓	–	–
	Short term care	✓	✓	–

Source own draft based on Long-term Care Insurance Blue Paper 2019

The assessment placed the eligible elders into different groups ranging 0–5 levels for functional disability (mainly on ADLs, level 3 is the higher level of partial dependent group and level 4 and 5 falls into total dependent group) and mental disability. Only level 3–5 functional disabled and MMSE $> = 9$ elders are eligible for LTCI subsidies.

6.2.4 Qingdao (2018) LTCI Benefit Standards

The new 2018 reform defines benefit structure as in Tables 6.2 and 6.3. Compared to the previous regulations, it is more comprehensive and co-pay ratios are more reasonable for residential members. The first four provision types shown in Table 6.2 are for ADLs based LTCI recipients and the latter three for mental disabled elders (usually more than 60 years old).

The comprehensive matrix tables for care subsidies form the current Qingdao LTC insurance benefit structure could deliver reasonable services to elders in need. Next section will turn to the cost analysis of the LTCI system.

---

[7] In 2018, EMIM contributes on average about RMB 3900 (EUR 548), both from employees (9% of wage) and individuals (2%). The RLMIM (including children) contributes RMB 260/year (EUR 36.50) compared to RHMIM of RMB 390 (EUR 55, university students counted as RHMIM paying only RMB 125, EUR 17.50), it is anticipated that most elders would choose RHMI programme (Qingdao Bendibao, 2019).

Table 6.2 Qingdao LTCI subsidy prices and individual co-pay ratios for integrated health care services

Provision type	EMIM	RHMIM	RLMIM
Hospital care	Top grade hospital RMB 210/day (EUR 29.50) Second grade hospital RMB 180/day (EUR 25) (RMB 300/day for tracheotomy patients, EUR 42)	–	
Institutional care	RMB 50/day (EUR 7)		–
Home care	RMB 50/day (EUR 7)	–	–
Mobile care	RMB 2500/year (EUR 351)	RMB 2200/year (EUR 309)	RMB 1500/year (EUR 211)
Dementia LTC	RMB 65/day (EUR 9)		–
Dementia day care	RMB 50/day (EUR 7)	–	–
Dementia short term care	RMB 65/day (EUR 9)		–
Individual co-pay (%)	10	20	30

Table 6.3 Qingdao LTCI subsidy prices for social LTC services for employee medical insurance members only

Provision type	Level 3 disability	Level 4 disability	Level 5 disability	Severely demented
Hospital care	–		RMB 1500/ month (EUR 211) RMB 50/day (EUR 7)	–
Institutional care	RMB 660/month (EUR 93) RMB 22/day (EUR 3)	RMB 1050/ month (EUR 147) RMB 35/day (EUR 5)		
Home care	3 h/week (RMB 50/h, EUR 7)	5 h/week (RMB 50/h, EUR 7)	7 h/week (RMB 50/hour, EUR 7)	
Mobile care				
Dementia LTC	–			RMB 1500/ month (EUR 147) RMB 50/day (EUR 7)
Dementia Day care				RMB 750/ month (EUR 105) RMB 50/day (EUR 7)
Individual co-pay	10%			

Table 6.4 Reports the general statistics of the Qingdao LTCI recipients during 2012–14 (Lu et al., 2017b)

ADLs Groups	0	5–15	20–35	40 +	All
ADLs distribution (%)	10	19	44	28	
Age < 60	170	235	478	398	5%
60–69	175	315	835	701	9%
70–79	473	933	2464	1826	24%
80–89	1029	2097	5047	3055	47%
90 +	443	861	1595	694	15%
Female (%)	63	60	62	61	61
Hospital care	529	600	837	452	10%
Home/Inst. care	1761	3841	9582	6222	90%
No. of recipients	2290	4441	10419	6674	23,824

6.3 Qingdao LTCI Cost Analysis

Since 2012, Qingdao has been operating the LTCI based on eligibility criteria which has been evolving and expanding. The key assessment criteria are ADLs scores which was previously set at 55 maximum but was extended to 60 to accommodate the needs of some dementia recipients in 2018, however, by far the majority eligible recipients were still below the score of 55. Mi et al. (2018) used the data which comprised 23,828 individual observations, covering the entire client activity of the Qingdao LTCMI from July 2012 through 15th April 2014. Within this period, 4,454 individuals exited the system when dired.

About 62% of beneficiaries are oldest old of age 80 and above, and 61% are females. About 10% received hospital long term care services. Majority are in home or nursing home care (with same subsidy, see Table 6.4).

6.3.1 Average Length of Stay in Qingdao LTCI (Mi et al., 2018)

The average entry ADLs score for Qingdao LTCI recipients was reported about 26.5 (using Barthel index 0–100, 100 stands for totally independent score). Simulation as made for the individual samples using multistate transition probability from home/institutional to hospital LT care.[8]

The simulation results indicated that recipients whose age is between 60–75 years, had similar length of stay of about 60 months, which declined dramatically for

---

[8] We ignore the mobile care as the sample size was very small during that period.

recipients at age 80 and above. Average length of stay for all LTCI recipients in Qingdao was 53 months. The length of stay for age 85 and above was 44 months.

6.3.2 LTCI Premium Estimations

The Shandong Bureau of Finance reported that between 2012 and 2016, the total expenditure amounted to RMB 1.13 billion (EUR 159 million). The Qingdao LTCI Blue Paper (Zhejiang University et al., 2018a, 2018b) provided a detailed breakdown of the insurance fund expenditure. This figure corresponds to the subsidy received by individual LTCI members for twice per week of services.[9]

This number enables the length of stay to be translated into a total cost of the system members based on the following:

Average individual cost C = Service Price PX length of stay T

$$C(agei) = \sum_{k=H,H,I} PT$$

All social insurance policies typically have predefined targets for benefit structure. In Qingdao's LTCI, approximately 10% of the recipients pertains to hospital care, with the reminder allocated to home/institutional packages. If individuals received services twice a week on average between 2012–2014, this would effectively double under the 2018 policy setting, given the inclusion of social care. An estimation of premium costs based on the recipients' age could be derived and is presented in Table 6.5 (Mi et al. 2018).

The cost analysis based on the current Qingdao LTC system operation offers a pioneering budgetary benchmark for long-term care cost in China. The average recipient in Qingdao in 2013 incurred costs equivalent to approximately 97% of the city's disposable income level. By 2018, with the inclusion of social care, costs were expected to double, providing services at a rate twice as frequent as in Scenarios 1. In Scenario 2, focusing solely on individuals aged 85 and above, individual cost decreased as the length of stay reduced. Scenario 3 assumes an assumption of increasing the ratio of high-care hospital coverage, resulting in a 22% cost increased for doubling the current 10% hospital care ratio, reflecting the effects of population ageing.

These figures serve as valuable reference for other regions in China to formulate LTC budgets. They can be translated into yearly costs by dividing the total cost by the length of stay. This can also be viewed as a baseline evaluation model for future system developments. As the disability ratio and quantity of service provisions change, so too will the system's costs.

Several issues are important for this cost analysis. First, the price of hospital care is about one third of a normal hospital bed in Qingdao. This is a market price

---

[9] Calculated by authors.

Table 6.5 Average Qingdao LTCI recipient's cost within the 2012–2014 system and scenarios-based cost analysis [10]

	2013 Qingdao	Scenario 1	Scenario 2	Scenario 3
	Recipient	Increase service	For age 85 +	Increase hospital ratio
Ave. total months stay	53	53	44	52
Hospital care ratio (%)	10	10	10	20
Ave. services per week	2	4	4	2
Cost with average service (RMB)	34,319 (EUR 4819)	68,638 (EUR 9638)	57,734 (EUR 8107)	41,869 (EUR 5879)
Full cost (RMB, 7 days/week)	120,117 (EUR 16,866)	120,117 (EUR 16,866)	101,034 (EUR 14,187)	146,542 (EUR 20,577)
Scenario cost (% of 2013 disposable income)	97	195	164	119

Source Mi et al. (2018)

where supply and demand have reached equilibrium based on years of Qingdao's experience. The same is true for the hourly charge of clinic care and community service for home care recipients. Second, the current eligibility criteria in Qingdao are largely based on ADLs and disease conditions, with costs derived from these statuses. For most new experiments in long-term care systems, defining eligibility and assessment criteria is challenging, as they directly influence not only the number of applications but also the average length of stay in the system. Qingdao LTCI provides valuable insights into these conditions for considering a LTCI system.

6.4 Qingdao LTCI Impact to Its Total Health Expenditures

One of the primary motivations behind establishing the LTC system is to balance healthcare resources by directing non-acute cases to long term care service provisions, such as LTC beds. Japan introduced its LTC insurance in 2000 and the total healthcare cost budget illustrated the substitutional effect of medical health costs by LTC insurance.

Following the implementation of LTC in 2000, medical expenditures in Japan experienced a significant drop and have since remained relatively stable. This provides another perspective for evaluating a LTC system: its ability to substitute

---

[10] Detailed models refer to Mi et al. (2018).

some medical expenditure while enhancing the welfare of recipient through improved care services. We aim at analyse Qingdao's performance based on this rationale.

Although sufficient aggregated data is not available to plot such a chart of substitutional effect, Lu et al. (2020) used individual data to analyse the Qingdao LTCI recipients' medical expenditure before and after entering the LTCI.

Recipients were grouped by 12 months expenditure before and after LTCI system. The expenditure was further subdivided into medical expenditure by the medical insurance fund and out of pocket by individuals. Table 6.6 summarised the 12 months performance before and after the Qingdao LTCI.

The table demonstrates a significant 167% reduction in total health expenditure after elders are admitted into the system. There was a 119% decrease in the public funding, comprising the total of medical insurance and long-term care insurance fund expenditure), and a total 48% reduction in out-of-pocket expenditures. The increase in long-term care expenditure is significantly smaller than the reduction in other medical and health care expenditures. Using a difference-in-difference model, Lu et al. (2020, p8) further analyzed the effect on medical cost before and after the implementation of the LTC system by controlling individual factors. They found that there is an average reduction of RMB 10,242 per recipient after they are admitted to the LTC system (comparing 12 months before and after), with RMB 2,324 coming from out-of-pocket reduction and RMB 7,918 from medical insurance expenditure.

This implies that the performance of Qingdao LTCI is efficient. Given that the Chinese health system heavily depends on major public hospitals, especially considering the nascent stage of primary care at clinic level in most regions, the implementation of Qingdao LTCI has effectively redistributed non-acute cases are distributed

Table 6.6 Qingdao LTCI recipients' 12 months health expenditures before and after entering the system, as a % of Qingdao city average disposable income in 2013

		Before	After	Difference (%)	significant
Total expenditure	Total	342%	175%	−167%	0.0000
	Inpatient	302%	91%	−211%	0.0000
	LTC	0%	76%	76%	0.0000
	Others	40%	8%	−32%	0.0000
Out-of-pocket	Total	84%	36%	−48%	0.0000
	Inpatient	74%	19%	−54%	0.0000
	LTC	0%	14%	14%	0.0000
	Others	10%	2%	−8%	0.0000
Medical insurance fund expenditure	Total	258%	139%	−119%	0.0000
	Inpatient	228%	72%	−156%	0.0000
	LTC	0%	61%	61%	0.0000
	Others	30%	6%	−24%	0.0000

Note: The table is taken from Lu et al. (2020, p4) using Table 2 numbers and divided by the 2012 Qingdao average disposable income per capita

to local clinics and community health care centres (as part of non-inpatients cost items). This has significantly contributed to the reform aimed at reallocating resources towards primary care structure. Additionally, it has been observed that LTCI has greatly reduced the financial burden on frail individuals, and their access to service provisions has became more convenient.

One of the most notable achievements of Qingdao LTCI is its success in reducing and containing average inpatient expenditures within the medical insurance fund. The inpatient costs for LTC eligible elders are less than a third compared to before their enrollment in LTCI. This contributes to alleviate pressure on major national hospitals, allowing them to allocate more resources to acute care cases. In addition, recipient receive better care through more frequent clinical visits within residential places (or at aged care hospital beds at a more economical cost). Qingdao's LTCI has played an important role in transiting the previous hospital-centered health care system into a multi-tier structure. This experience is extremely valuable in China's current context, where all medical resources are concentrated in major national hospitals. Qingdao LTCI has served an effective example of how to reform this system.

6.5   Challenges, Implications and Conclusions

Though the Qingdao LTCI is one of the best pilot programs in China, some outstanding issues are yet to be reformed and there are still rooms for future accomplishment of a stellar LTC system.

6.5.1   Equality and Integration

Under the 2018 policy structure, residents (non employee pensioners) are not entitled to certain services, such as home care and social care. The Bureau of Civil Affairs is responsible for financially disadvantaged residents who are not covered by LTCI services. Although integrating these two systems is challenging, better results could be achieved by incorporating the services into one policy framework to improve efficiency. This is partly due to the current overlap in China's administrative systems. China has implemented dual social insurance systems for almost all welfare services, separating formal employees from other residents. They are usually under different administrative entities. A system integration is needed at a higher level. It is worth noting that greater equality can be achieved under a unified policy administration.

6.5.2 Future Demand for Dementia Care

Dementia care is the last step to be integrated into the Qingdao LTCI after careful considerations. Although at this stage the proportion of dementia recipients is still small, statistics from developed countries have shown that numbers could increase to about half of all recipients as the system matures. It would be prudent for the government to establish an intervention fund to manage citizens with cognitive decline citizens, however, it remains unclear how this fund will be allocated and how effective this would be. Models of dementia care might be imperative as the degree of ageing intensifies.

The Qingdao LTCI pilot program provides an integrated and comprehensive healthcare system that addresses the needs of an ageing society at its current stage. The evolution of policy reforms carries important implications for other regions in China that are in the initial stages of planning a long-term care policy planning. This empirical analysis of the Qingdao program provides valuable insights into the cost mechanisms at both individual and system levels, generating overall budgetary estimations.

Acknowledgements The project is funded by ARC Grant LP150100347 in Australia, and is supported by the ARC Centre of Excellence in Population Ageing Research (CEPAR), ARC grant number CE11E0099. This research is also supported by a Major Project of the National Nature Science Foundation of China (grant number 71490733).

References

Cheng, Y., & Shen, Y. J. (2017). China long-term care pilot projects comparison and discussions: Based on five regional policy analysis. *Public Governance Review, 001*(2017), 15–24.

Deng, D. S., & Guo, T. (2015). The construction of China's long-term care insurance system: A case study of Qingdao. *Research on Health Economics, 10*(342), 33–37.

Du, P., & Wang, Y. M. (2016). Population ageing and the development of social care service systems for older persons in China. *International Journal on Ageing in Developing Countries, 1*(1), 40–52.

Gu, D. N., & Zeng, Y. (2006). Changes of disability in activities of daily living among Chinese elderly from 1992–2002. *Population and Economics, 4*, 9–13.

Gu, D., Dupre, M., Sautter, J., Zhu, H., Liu, Y. Z., & Zeng, Y. (2009). Frailty and mortality among Chinese at advanced ages. *Journal of Gerontology: Social Sciences, 64b*(2):279–289.

Hu, H. W., Li, Y. N., & Zhang, L. (2015). Estimation and prediction of demand of Chinese elderly long-term care service. *Chinese Journal of Population Science, 3*, 79–89.

Jin, T. (2006). *Long term care insurance: A very competitive insurance product in future China.* China Foreign Economics and Trade University Publishing House.

Lin, B. (2015). Preliminary study on the model of long-term care insurance in China. *Scientific Research on Ageing, 3*(5), 13–21.

Lu, B., Liu, X., & Piggott, J. (2015). Informal long-term care in China and population ageing: Evidence and policy implications. *Population Review, 54*(2), 28–41.

Lu, B., Liu, X. T., & Yang, M. X. (2017a). A budget proposal for China's public long-term care policy. *Journal of Ageing and Social Policy, 29*(1), 84–103.

Lu, B., Piggott, J., Zhu, Y., & Mi, H. (2017b). A sustainable long-term health care system for ageing China: A case study of regional practice. *Health Systems and Reform, 3*(3), 1–9.

Lu, B., Mi, H., Yan, G., Lim, J. K., & Feng, G. (2020). Substitutional effect of long-term care to hospital inpatient care? *China Economic Review, 62*, 101466.

Ma, J., Zhu, M. L., Xiao, M. Z., & Song, Z. J. (2012). *China health expenditure and estimation of fiscal pressure, China national balance account studies.* Social Science Publishing House.

Mi, H., Lu, B., Fan, X. D., Cai, L. M., & Piggott, J. (2018). Preparing for population ageing: Estimating the cost of formal aged care in China. *Journal of Economics of Ageing, Forthcoming.*

Qingdao Bendibao. (2019). Qingdao medical insurance payment. Retrieved from December 12, 2020: http://qd.bendibao.com/live/20171020/51233.shtm

Qingdao Daily. (2016). Low wages, low status, large gap of care workers in elderly care in Qingdao (24.03.2016). Retrieved from May 23, 2017: http://www.dailyqd.com/news/2016-03/24/content_319318.htm

Qingdao Municipal Peoples Government. (2018). Qingdao municipal peoples government gazette (Doc. No. 3, 4, 5 and 12). Retrieved from December 12, 2020: http://www.chengyang.gov.cn/n1/upload/181211155739701422/181211155739251311.pdf

Wei, H. L., & He, Y. D. (2012). Research on potential insurance market for China's long-term care. *Insurance Studies, 7*(2012), 7–15.

Willemé, P. (2010). The Belgian long-term care system. Retrieved from December 12, 2020: https://www.plan.be/uploaded/documents/201004230943350.wp2001007.pdf

World Bank. (2016). *Living long and prosper: Ageing in East Asia and Pacific.* Washington, DC: World Bank East Asia and Pacifica Regional Report, International Bank for Reconstruction and Development. Retrieved from December 12, 2020: https://pubdocs.worldbank.org/en/632851464598111066/053116-ageing-EAP-Philip-Okeefe.pdf

Yang, W., He, A. J. W., Fang, L. J., & Mossialos, E. (2016). Financing institutional long-term care for the elderly in China: A policy evaluation of new models. *Health Policy and Planning, Journal on Health Policy and System Research, 31*(10), 1391–1401.

Zhejiang University; Qingdao Human Resources and Social Security Bureau and Qingdao Finance Bureau (Eds.). (2018a). Long-term care blue paper 2018: Qingdao long-term care insurance studies.

Zhejiang University; Qingdao Human Resources and Social Security Bureau and Qingdao Finance Bureau. (2018b). Long-term care insurance blue paper: Research on long term care insurance in Qingdao.

Zhu, M. L., & Jia, Q. X. (2009). The analysis of demand for long term care and its insurance system constructing in China. *Chinese Journal of Health Policy, 2*(7), 32–38.

Bei Lu is a Senior Research Fellow at the Centre of Excellence in Population Ageing Research (CEPAR), which is in the Australian School of Business at the University of New South Wales. She is also a Research Fellow at Zhejiang University and Tsinghua University. She been focusing on pension economics, long-term care systems and healthcare research, especially in China. She assisted the World Bank's China pension projects in 2004 and 2016, and United Nations' project of population ageing and fiscal sustainability in East Asia and Pacific in 2017. Her research has appeared in the Health and Social Care in Community, Journal of Ageing and Social Policy, International Social Security Review, the Journal of the Economics of Ageing, Population Review, and CeSifo Economic Studies. She has also pub-lished several articles in Chinese books, newspapers and journals.

Guanggang Feng is an Associate Professor at the Department of Labour and Social Security, School of Finance and Public Administration, Anhui University of Finance and Economics. He was awarded his PhD. in Economics from the Shanghai University of Finance and Economics. His recent research focuses on long-term care empirical studies, pension actuarial and economic demography.

Part III
Care Education

Chapter 7
Curriculum Development in Nursing Education in Germany

Ingrid Darmann-Finck

7.1 From Special Educational Programmes to General Nursing Education—Nursing Education in Germany

After a long time of political dispute, the German parliament passed a reformed law for the nursing profession in 2017. In this law, a new professional profile was established in Germany: the general nurse. In this contribution, the background of the reform is described. Besides this new qualification, two of the former vocational qualifications—the nurse for the elderly and the paediatric nurse will remain. But all students, independent of the above qualifications they pursue, must undertake the first two years of the educational programme together, which focuses on general nursing. The law came in force in 2020. The German nursing schools—where most of the nurses in Germany get their professional licensing by a vocational education—are now faced with the task to develop new curricula.

Usually, curricula are developed very pragmatically. Below, the basics of a curriculum development based on educational assumptions are presented and visualised by some examples.

In Germany, until now there have been three nursing professions. The specialisation of each profession is based on the age of the target groups. There is a nursing profession for paediatric nursing, one for elderly care and one for sick people, which focuses on the stages of life between youth and old age and on nursing in hospitals. This differentiation has primarily historical reasons. The historically oldest nursing profession is the nurse for sick people. In 1781, for the first time an educational programme was established. During this period, medical science expanded, and the first modern hospitals were founded. The first professional law for nursing education and a state examination was adopted by an Act of Parliament in 1907 (Kruse,

I. Darmann-Finck (✉)
University of Bremen, Bremen, Germany
e-mail: darmann@uni-bremen.de

Institute for Public Health and Nursing Research at University of Bremen, Bremen, Germany

© The Author(s) 2025
Deutsche Gesellschaft für Internationale Zusammenarbeit (GIZ) GmbH et al., (eds.),
Sustainable Aging, https://doi.org/10.1007/978-3-662-69139-7_7

1987). The next profession which arose, was paediatric nursing. An educational programme with a focus on nursing for babies and children is known since the end of the nineteenth century. The initiative came from a paediatrician, for him it was one step to reduce the high infant mortality in this period. A first professional law and state examination for paediatric nurses, was legally anchored in 1917. After the Second World War, in the 1950s, there was a discussion on the improvement of the care for older people who lived in institutional homes in Germany (Sahmel, 2001). Until then, in homes, often only the home management had a professional education, most of the other workers didn't have any nursing educational background. Furthermore, the number of old people who needed institutional care grew. Two reasons are responsible for this: the growing number of old and very old people and the change of family structures. For example, because of the increasing number of working women, younger generations couldn't care for the older generations like in former times. At the end of the 1950s, first small educational programmes were established. In 1969, for the first time a legal regulation for the education of the elderly care was adopted in one federal state of Germany (Sahmel, 2001). Since 2003, there is a federal law about the professional education of nurses for elderly care.

For nearly twenty years, there have been discussions about a generalisation of nursing education, because the care requirements in the different sectors of nursing care have changed based on the demographic and epidemiological developments. In hospitals, for example, in all units on average, 20% of patients suffer from dementia as a secondary diagnosis (Isfort et al., 2014: 38). That means that nurses in hospitals don't only need competencies to care for people with acute diseases, they need competencies to care for the elderly too. On the other hand, residents of nursing homes aren't only old, but most of them are sick too. 80% of them need for example medical-diagnostic or medical-therapeutic performances to an extent of 7 h a week on an average (TNS Infratest Sozialforschung, 2017: 313). For this reason, nurses in long-term care need competencies to care for sick people too. And finally, paediatric nurses need competencies for the care of adults also. Children and juveniles with chronic diseases Cowley grow up and the nurses should be able to lead the juveniles over to the care of adults (Cowley et al., 2011). In addition, in the next 30 years when the baby boomers in Germany become old there will be a growing need for care but limited resources of nursing care. Healthcare researchers concluded that this situation can only be managed if old people stay if possible at home, and if relatives and voluntary workers have greater involvement in the care of older people (Rothgang et al., 2012). Also, the tasks of nurses have to change: today nurses need to focus on care and case management, consultancy and effective interface management. This suggests that nurses should not only require competencies for direct care but also systemic competencies. Besides the arguments deduced from care requirements, there are further arguments against the existing iniquity between the three professions. The elderly care nursing profession has a bad image and the salary of the nurses in long-term care is much lower than the salary of nurses in hospitals. Against this background, most of the German nursing researchers and the nursing associations' demand that the three different nursing professions and the three educational programmes should be joined together (BMFSFJ, 2008; DBR,

2006). In 2016 the two competent federal ministries, the Federal Ministry of Health and the Federal Ministry of Family Affairs, submitted a draft legislation in which the division of nursing into three professions was abolished, and a common education for a general nurse was established. But because of political resistance in 2017, the government coalition found a compromise: now all students learn together for the first two years. After this, some students (probably around 50%) can take up specialisation in elderly care or for children and juveniles in the last year. The rest of the students will finish the education for general nurses. The exclusive course on nursing for sick people doesn't exist anymore.

7.2 Curriculum Development in Germany

There are different methodical approaches of curriculum development with different educational and scientific assumptions. Internationally the six-step approach has prevailed, which was devised by Kern et al. (2009) at John Hopkins University of Baltimore (USA). The approach starts with a (i) problem identification and general needs assessment, and a (ii) targeted needs assessment. With the general needs assessment deficits in health care are determined, and then approaches are identified which can serve as a basis on how problems should ideally be addressed. Methods for obtaining necessary information are a literature review or consultations with experts. The targeted needs assessment aims at the special needs of the learners by comparing the ideal and the actual characteristics of the targeted learner group. Building up on the results, (iii) goals and objectives are defined and then (iv) educational strategies can be planned. The final stages are (v) implementation of the curriculum and (vi) evaluation and feedback. In contrast to the education-based approach, the six-step approach has no educational goals and focuses only on functional requirements of the nursing practice.

That is why in Germany, often an approach is preferred which is based on educational theories (Knigge-Demal, 2001; Siebert, 1974). This approach aims at critical personality development and increasing emancipation. It intends to reach educational goals besides functional qualifications. In concrete terms, for instance, students learn to reflect on the contradictions which relate to nursing practice. The contradictions may refer to internal conflicts or to institutional contradictions. For example, the contradiction of standardisation of nursing on the one hand and individualised care on the other hand or the contradiction between the principles of caring and autonomy. These critical-reflective action competences are essential because of the power potential of nurses and the dependency of persons being cared for. Furthermore, often deficits in health care cannot only be resolved by educational programmes. They need fundamental changes in the health care system. Therefore, educational programmes should enable the students to recognise and analyse such problems and to actively shape professional changes. In addition, from an educational point of view, an initial training should always relate to educational goals.

The following steps belong to the education-based approach (see Fig. 7.1). In the first step, the participants of a curriculum development working group have to decide the educational framework of the curriculum. This framework should include overarching objectives of the educational programme, such as general goals of personality development and goals concerning the professional role and the ethical fundament of nursing. For example, different nursing theories might be used as a theoretical fundament of the curriculum. From the point of view of critical educational theory, a critical nursing theory should be selected because only on this basis, the students can build on critical and reflexive competences.

In the second step the members of a curriculum developing group have to proceed different preparatory analysis similar to the approach of Kern. In this approach, they also have to find out the actual and the expected prospective qualification requirements in nursing practice. Based on qualification research, the situations and tasks could be identified for which nurses must be qualified. Often used in international contexts is DACUM (Developing A Curriculum) Method for identification and analysis of professional work tasks (Norton, 1997). This method is criticised because it sets the decontextualised tasks as a starting point of curriculum development. In Germany, this approach was further developed by an occupation-educational research perspective. From this perspective, characteristic work situations are identified by empirical research (e.g. by action-oriented specialised interviews, expert

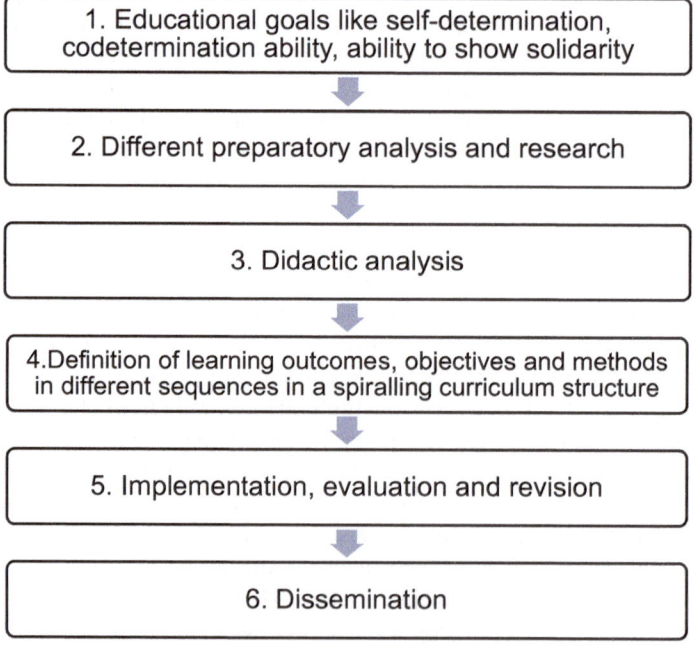

Fig. 7.1 Phases of the educational based method of curriculum development. *Source* Darmann-Finck

skilled worker workshops or non-participative observation), and later these are structured according to developmental logic based on theories of competence development (Pahl & Rauner, 2008). In the field of nursing, it is further insufficient to determine only situations that could be resolved with knowledge of rules. It's necessary to collect conflict and dilemma situations as well because nursing practice is characterised by unawareness, of little potential for standardisation and contradictory requirements. Only based on such situations, analysed under the perspective of competence development, students can learn to reflect and to deal with the nursing dilemmas and contradictions, and to avoid one-sided solutions (Darmann-Finck, 2014; Darmann-Fink et al., 2008). In parallel, the curriculum working group has to conduct a comprehensive literature research, so that they are informed about the actual scientific knowledge in nursing science. On the basis of this research, theoretical knowledge and action patterns could be identified, which can be used to understand the nursing situations found by qualification research and to develop alternatives of action. Finally, it is part of the preparatory analyses to assess the learning conditions of the students regarding the competencies they have to obtain.

In the next phase of construction, the findings of the preparatory analysis are analysed by educational categories. For this analysis, different educational models can be used. These models supply categories to systemise the findings under an educational perspective.

Afterwards the real construction of the curriculum takes place. This step is explained in detail in the next chapter.

A curriculum is an important basis for the control of educational programmes, but it can't determine the implementation by teachers. The implementation process must be planned and executed carefully. Hindering and beneficial factors that affect the implementation must be taken in consideration. The evaluation phase is a very important component of the process of curriculum development and implementation too, because the results show if the curriculum is effective in reaching the intended learning outcomes (intern validity). Furthermore, the extern validity, the actuality of the curriculum should be monitored regularly. Consequences could be drawn from the results for changes and upgrades.

7.3 Principles of Curriculum Construction

In vocational education in Germany, the situation-based principle of curriculum construction combined with outcome (competency) orientation has prevailed in the last decades (Reetz & Seyd, 2006). Situation-based means that curricula are not structured by the systematic of sciences, but by professional tasks and situations nurses have to cope in nursing practice. Starting from this situation, students require knowledge and competences. With this curriculum structure, there are beneficial conditions for case-based and action-oriented learning and this learning approach is very suitable to foster action competence (CTGV, 1990). Because students gain competences in the context of the situations in which the competences are needed in practice, the learning is more application-related than before. In contrast, a curriculum based on the systematic structure of sciences is frequently associated with receptive learning and the accumulation of inert knowledge.

A situation-based curriculum section involves an interdisciplinary approach to different sciences. But within a situation-based curriculum section, not all the lessons are necessarily case-based. Within the overarching units too, lessons that intend a systematic acquisition of knowledge or that focus on the development of personality by acquisition of (formal) competences e.g. empathy trainings, can be integrated.

Based on increasingly complex situations, students are able to acquire nursing competence (Benner, 1984). Benner describes the acquisition and development of clinical competence in five stages, but only the first three stages are relevant for initial nursing education. At the beginning of the clinical education, the behaviour in the clinical setting is very inflexible. Students act along rules and solution schemes. In the next stage, students are increasingly able to recognise recurrent, meaningful components of situations and to adjust slightly, the used action patterns. At the end of their education, students should develop the competence to recognise patterns and the nature of clinical situations more quickly and to modify plans in response to the complex influencing factors. But in this stage, they still find solutions based on analytical thinking. After initial training and some years of experience, nurses are more and more able to understand situations as whole parts and to get an intuitive judgement of clinical situations (Benner, 1984).

The gradual build-up of competencies in this sense is fostered by increasingly complex situations and requirements as learning opportunities in a nursing education curriculum.

- Simple requirements: They could be resolved by simple cause-effect-patterns and lead to a predictable result.
- Complicated requirements: In such situations, some or a lot of complicated conditions exist. To find a resolution, the sum of these factors has to be calculated, but there is still a predictable result.
- Complex requirements are characterised by many different influencing factors and their relations. The influencing factors as well as their interactions are uncertain and the solution process can't be carried out in a linear mode. Predicting the resolution is therefore difficult and only possible within limits (Benner, 1984; Rauner, 1999). In the interactionist nursing didactics, a German nursing didactic model, so called key problems are integrated in the curriculum (Darmann-Finck, 2010). Key problems have particularly high educational potential because they have very different solutions which are defined by divergent interests and values. They must be addressed in a multidimensional, interdisciplinary manner and are rooted in structurally defined professional antinomies or contradictions, as described before. The aim of nurse training is to make students aware of these contradictions, and by reflecting on these, avoid one-sided solutions.

To create situations in the context of curriculum development four characteristics of situations by Kaiser (1985) can be used. These are: action patterns and their theoretical and empirical reasons; the role structure, the purpose of the situation as well as the setting of the situation.

The situational structure of the curriculum can be developed by combining the four characteristics of a situation in a sensible way. For example, the action patterns health promotion and health counselling or education could be combined with a chronic disease like diabetes mellitus as purpose of the situation and perhaps a patient who does not want to adapt to meet rigorous rules. As a situational setting the ambulant care could be selected. As another example, the action pattern transition from inpatient to outpatient care must be combined with a complex disease which requires a lot of managing activities, such as a stroke or an oncological illness with need of care. The situational setting consists of the interface between hospital and ambulant care. The role structure can be very complicated, since different professions and different members of the family or other related persons must be integrated.

The complexity of situations can be reached by using complicated action patterns and by a careful selection of the other characteristics of situations. For example, a situation becomes much more complicated if the patients suffer from complex diseases, if special cultural aspects need to be taken into consideration or if there are close economic limits in the situational setting.

7.4 Conclusion

In this contribution, an education-based approach of curriculum development is presented. In fact, curriculum working groups often proceed very pragmatically and do not realise an in-depth empirical and theoretical analysis. The resulting curriculums often don't meet scientific and educational criteria. They don't intend a forward-looking professional profile and maintain the status quo.

In the last few years in Germany, the care of the elderly became much more the care of sick elderly people, so that the profession which formerly had a social orientation developed a more medical focus. That is why in Germany, most of the nursing researchers and the nurse associations argue that the general nurse profile is the right resolution for future health problems. Other countries which have to build up long-term care facilities can't be recommended to create a special nursing profession for elderly care. Instead, the profile of the general nurse should be enlarged by elderly care aspects. In any case, in elderly care a qualification mix with registered nurses and nursing aids, with and without special education is necessary.

References

Benner, P. (1984). *From novice to expert, excellence and power in clinical nursing practice*. Addison-Wesley Publishing Company.

BMFSFJ, Bundesministerium für Familie, Senioren, Frauen und Jugend. (2008). *Pflegeausbildung in Bewegung*. Ein Modellvorhaben zur Weiterentwicklung der Pflegeberufe. Schlussbericht der wissenschaftlichen Begleitung. Retrieved from: https://www.dip.de/fileadmin/data/pdf/material/PiB_Abschlussbericht.pdf, June 22, 2018.

CTGV, Cognition and Technology Group of Vanderbilt. (1990). Anchored instruction and its relationship to situated cognition. *Educational Researcher, 19*(6), 2–10.

Cowley, R., Wolfe, I., & McKee, M. (2011). *Improving the transition between paediatric and adult healthcare: a systematic review*. Retrieved from: https://pdfs.semanticscholar.org/bd28/b1085559ec5a3261e5d84451dfa55e9d773c.pdf, June 22, 2018.

Darmann-Finck, I. (2010). Interaktion im Pflegeunterricht, Frankfurt a. M.: Lang.

Darmann-Finck, I. (2014). Berufswissenschaftliche Erhebungen als Grundlage für die Curriculumentwicklung unter pflegedidaktischer Perspektive. In G. Spöttl, M. Becker, & M. Fischer (Eds.), *Arbeitsforschung und berufliches Lernen* (pp. 263–277). Lang.

Darmann-Finck, I., Keuchel, R., & Myrick, F. (2008). Health/Care. In F. Rauner & R. Maclean (Eds.), *Handbook of technical and vocational education and training research* (pp. 227–233). Springer.

DBR, Deutscher Bildungsrat für Pflegeberufe (Ed.). (2006). *Pflegebildung offensiv*. Elsevier, Urban & Fischer.

Isfort, M., Klostermann, J., Gehlen, D., & Siegling, B. (2014). *Pflege-Thermometer 2014. Eine bundesweite Befragung von leitenden Pflegekräften zur Pflege und Patientenversorgung von Menschen mit Demenz im Krankenhaus*. Herausgegeben von: Deutsches Institut für angewandte Pflegeforschung e.V. (dip), Köln. Retrieved from: http://www.dip.de/fileadmin/data/pdf/projekte/Pflege-Thermometer_2014.pdf, June 22, 2018

Kaiser, A. (1985). *Sinn und Situation. Grundlinien einer Didaktik der Erwachsenenbildung*. Klinkhardt.

Kern, D. E., Thomas, P. A., & Hughes, M. T. (2009). *Curriculum development for medical education: A six-step approach* (2nd ed.). Johns Hopkins University Press.

Knigge-Demal, B. (2001). Curricula und deren Bedeutung für die Ausbildung. In M. Sieger (Eds.), *Pflegepädagogik. Handbuch zur pflegeberuflichen Bildung* (pp 39–55). Huber.

Kruse, A.-P. (1987). *Die Krankenpflegeausbildung seit der Mitte des 19. Jahrhunderts*. Kohlhammer.

Norton, R.E. (1997). *DACUM Handbook*. The National Centre on Education and Training for Employment. State University.

Pahl, J.-P., & Rauner, F. (2008). Research in the vocational disciplines. In F. Rauner & R. Maclean (Eds.), *Handbook of technical and vocational education and training research* (pp. 193–199). Springer.

Rauner, F. (1999). Entwicklungslogisch strukturierte berufliche Curricula: Vom Neuling zur reflektierten Meisterschaft. *Zeitschrift für Berufs- und Wirtschaftspädagogik, 95*(3): 424–446.

Reetz, L., & Seyd, W. (2006). Curriculare Strukturen beruflicher Bildung. In R. Arnold & A. Lipsmeier (Eds.), *Handbuch der Berufsbildung* (2nd ed., pp. 203–219). VS Verlag für Sozialwissenschaften.

Rothgang, H., Müller, R., & Unger, R. (2012). *Themenreport „Pflege 2030". Was ist zu erwarten—was ist zu tun?* Bertelsmann Stiftung. Retrieved from: https://www.bertelsmann-stiftung.de/fileadmin/files/BSt/Publikationen/GrauePublikationen/GP_Themenreport_Pflege_2030.pdf, June 22 2020

Sahmel, K.-H. (2001). Bestandsaufnahme Pflegeausbildung. In K.-H. Sahmel (Ed.), *Grundfragen der Pflegepädagogik* (pp. 30–64). Kohlhammer.

Siebert, H. (1974). *Curricula für die Erwachsenenbildung*. Westermann.

TNS Infratest Sozialforschung. (2017). Abschlussbericht. Studie zur Auswirkung des Pflege-Neuausrichtungs-Gesetzes (PNG) und des ersten Pflegestärkungsgesetzes (PSG). Retrieved from: https://www.bundesgesundheitsministerium.de/fileadmin/Dateien/5_Publikationen/Pflege/Berichte/Abschlussbericht_Evaluation_PNG_PSG_I.pdf, June 22, 2018

Ingrid Darmann-Finck has worked as a Professor of Nursing at the University of Bremen's Institute for Public Health and Nursing Research since 2003, where she specialises in nursing education. She holds a Ph.D. from the University of Hamburg and has also trained in nursing and teacher training. She previously worked as a nurse from 1988 to1998 as well as a research fellow at the University of Hamburg from 1994 to 2003.

Chapter 8
Professional Competence of Teachers in the Training of Nursing Specialists Against the Background of Practice-Related Competence Models

Astrid Seltrecht

8.1 Education Export from Germany to China in the Field of Nursing Teacher Training

In Germany, the nursing profession has been regulated since 2020 by a Nursing Professions Act which was passed in 2017. The Nursing Professions Act also led to a reform of the nursing training programme. During this time of the training reform of the nursing professions in Germany, a Chinese college asked in 2017 to install the German training system for geriatric nursing, which was still in place at the time but was foreseeably coming to an end, at the Technical and Vocational College in the city of Panjin/China with German support. In the project 'Promotion of dual training for geriatric nurses at the Technical and Vocational College of the city of Panjin/China—educational analysis and curriculum development' (Seltrecht 5/2017–9/2017), there was a request from the Chinese side for a dual vocational training programme based on a curriculum that enables action-oriented and competence-oriented teaching. This curriculum was supposed to be developed with a group of selected teachers from the local college. The follow-up project 'Promotion of dual training at the college in the city of Panjin/China—teacher training for the introduction of a new curriculum' (Seltrecht 10/2017–1/2018) concentrated on training teachers to introduce and implement the new curriculum.

The education export from Germany to China that was planned for the project initially included an analysis of the previous training in nursing at the Panjin Technical and Vocational College. The aim was to design professional training in the field of 'nursing with specialisation in geriatric care' on the basis of this training course analysis and to develop a corresponding curriculum for this. For this purpose, the

A. Seltrecht (✉)
Faculty of Humanities, University of Magdeburg, Magdeburg, Germany
e-mail: Astrid.Seltrecht@ovgu.de

© The Author(s) 2025
Deutsche Gesellschaft für Internationale Zusammenarbeit (GIZ) GmbH et al., (eds.),
Sustainable Aging, https://doi.org/10.1007/978-3-662-69139-7_8

teachers were to be instructed and intensively involved—on the one hand, in order to incorporate framework conditions such as legal, economic and cultural regulations and traditions, particularly at the level of the college, the province and society as a whole. On the other hand, the intensive involvement was based on the professional assessment of how a jointly developed curriculum with teachers can subsequently be more easily implemented and, if necessary, modified by these teachers. The teachers who were involved in the development of the curriculum were to act as multipliers in the introduction and implementation of the curriculum and were to be supported in this new role by the German side.

In order to carry out the project, an analysis of the conditions was necessary. To perform this analysis of conditions, criteria were needed to enable a comparison of the two countries. The following concepts or categories were selected, which are anchored either in educational practice, educational policy or scientific theory:

- *double case and system reference* as a unique feature of the teaching activity in person-centred disciplines, which is anchored in the meta-professional teaching and research concept in the training of aspiring nursing teachers,
- *regulated and non-regulated professions* as a category to examine the possibility of introducing a new profession only at a college,
- *Competence development and competence models* to determine the flexibility and simultaneous reliability of the core category of a curriculum.

The examination of these three perspectives reveals differences between the two countries on the one hand, but also shows—particularly with regard to Germany— areas that need to be focussed on more closely in the future.

8.2   The Double Case and System Reference

The discussion of the professional competence of teachers involved in the training of aspiring nursing specialists inevitably leads to a discussion of the "double case and system reference". The concept of the double case reference is based on professional theoretical concepts and theories. It states that, for one, teachers must understand, reconstruct and evaluate the nursing-specific case relationships that exist between the nursing personnel and the patient in a theory-based manner for teaching purposes. However, as a teacher, you do not necessarily have to be capable of applying practical nursing techniques on a person (patient, resident) requiring nursing care. The theoretical nursing-specific case references are part of the nursing training programme, as they help future nursing specialists to acquire professional competence. For teaching in vocational and higher education nursing training, teachers construct professional scenarios or didactically prepare existing scenarios for this purpose. Occasionally, however, nursing-specific case descriptions are introduced spontaneously into the lessons by trainees. In these situations, the teacher is required to give a nursing-scientific interpretation of this case description. But the teacher must also decide pedagogically in which form this case should be included in the current teaching situation: Should it become the direct subject of the current lesson? Should the teacher discuss the case with the student in a face-to-face meeting after the lesson? Should the case description be addressed in a future lesson? Or does a case presented by a trainee remain without comment and is not taken into consideration? In this case, however, the nursing student would not receive a solution for the described problem and would be confused again in a comparable situation with a patient. The theoretical examination of a care practice would thus be avoided. In this moment—while the teacher is deciding how to deal with the example from nursing practice brought in by a student in the classroom—he or she is already in the pedagogical case reference: On an educational level, teachers must keep an eye on the processes that occur in the classroom and work on the constantly occurring, irresolvable antinomies. Teachers must therefore constantly keep both case references in mind in their work: the nursing-specific case reference and the educational case reference. A comparison of the two case references, however, reveals a difference between the two mentioned case references: Teachers must be able to theoretically penetrate, analyse and evaluate the nursing-specific case references. They must also be able to theoretically understand, analyse and evaluate the educational case references. In addition, teachers are also required to work on the educational case references in practice. In contrast to the nursing-specific case references, they need practical know-how in order to act appropriately towards the target group.

However, teachers do not only need to be familiar with the micro level of the two case references. They must also understand the rules and laws that apply in the various institutions and the existing structures in which the case references come into effect. Thus, teachers must also know two system references and consider them in their work: the relation to the vocational school or college system in which they are teaching, as well as the relation of the nursing personnel to the various nursing settings, e.g. hospital, nursing home, outpatient nursing service. The concept of the double case and system reference therefore focuses on the case-related, subject-orientated actions of teachers both in relation to the teaching and learning location of the vocational school or college and to the working and learning location of the nursing practice.

If the perspective is now shifted away from the nursing teachers and towards the lecturers who train the nursing teachers, it becomes clear that the double case and system reference is subject matter in the training of teachers. Lecturers at universities are themselves subject to a further case reference, namely that which exists between lecturers and students. For this reason, we speak of a 'triple case and system reference' in the context of teacher training. For this reason, the teaching and research concept at Otto von Guericke University Magdeburg, Germany, is based on the "triple case and system reference" (Seltrecht, 2015, S. 223). Since the teaching and research concept incorporates all system and case references across three professional fields of action, it is also referred to as a "meta-professional teaching and research concept" (Cf. Seltrecht, 2024).

8.3 Regulated and Non-regulated Professions in the Fields of Nursing Practice, Nursing Education Practice, Nursing Teacher Training Practice

8.3.1 Regulated and Non-regulated Professions in Germany

A comparison of the three practice areas of nursing, nursing education and nursing teacher training with regard to their professional admission requirements clearly shows that the three practice areas differ in terms of professional regulation. In Germany there is a distinction between regulated and non-regulated professions. Regulated professions are characterised by statutory regulations that define the qualifications and competencies necessary for practising a profession. Without these mandatory requirements, no licence to practise the profession will be issued. For the nursing profession and for the teaching profession, there are laws and regulations which govern the practice of the profession and the required competences. These regulations are, however, at different levels: at federal or state level. In contrast to nursing practice and nursing education practice, professional requirements for nursing teacher training practice are regulated by law for only a small group of lecturers.

8.3.2 Federally Regulated Profession: Nursing Specialist in Germany

In Germany, professional nursing care is regulated by the Nursing Professions Act (PflBG), which was passed in 2017 and came into force in 2020. The Act incorporates nursing care-specific and social changes: Demographic change, progress in nursing science and medicine, general and profession-specific digitalisation all have an impact on the challenges facing the nursing profession. The Nursing Professions Act therefore reorganises the nursing profession and at the same time reforms nursing training. 'Pflegefachfrauen' (female nursing specialists) and 'Pflegefachmänner' (male nursing specialists)—theese are the new job titles—work with nursing recipients of different ages (infants, children, adults, elderly people) in all areas of nursing care (acute inpatient, long-term inpatient, day-care centre, outpatient).[1] The Act stipulates in § 1 PflBG that a professional licence is required in Germany for working as a nursing specialist: 'Anyone wishing to use the professional title 'Pflegefachfrau' or 'Pflegefachmann' requires a licence.' Requirements for the issuance of

---

[1] In addition to the vocational qualification Pflegefachfrau/Pflegefachmann, the qualifications of healthcare and paediatric nurse and geriatric nurse still exist. Trainees can decide at the end of the second year of training which qualification they want to pursue in order to be trained specifically for this qualification in the third year of training.

the licence are regulated in § 2: "The authorisation to use the professional title shall be granted upon application if the applicant.

1. has completed the vocational or higher education training prescribed by this Act and has passed the state final examination,
2. has not committed any behaviour that would render him/her unreliable for exercising the profession,
3. is not unfit in terms of health to practise the profession and
4. has sufficient knowledge of the German language to practise the profession."(§2 PflBG)

Only persons who have been issued a professional licence are permitted to carry out the following activities in Germany:

1. the assessment and determination of individual nursing needs (…),
2. the organisation, arrangement and control of the nursing process (…) as well as
3. the analysis, evaluation, assurance and development of the quality of nursing care (…)' (§ 4 PflBG).

The training itself then takes place on the basis of a federal state curriculum that complies with the statutory regulations as set out in the Nursing Professions Act and in the training and examination regulations for the nursing professions.

8.3.3   Federal and State-Regulated Profession: Nursing Teacher in Germany

The task of high-quality vocational training is to prepare aspiring nursing specialists adequately for professional situations in nursing settings. At the federal level, that is, for Germany, the Nursing Professions Act regulates the admission requirements for teaching, differentiated according to whether theoretical or practical education is provided at a state, state-approved or state-recognised nursing school. This means that nursing schools must provide proof that they have professionally and pedagogically qualified teachers with a Master's degree (or equivalent) in nursing education for theoretical teaching. In order to carry out practical lessons, teachers with a nursing education degree (i.e. at least a Bachelor's degree) must be employed (see § 9 PflBG).

The requirements that must be met are assessed at federal state level and, if necessary, further requirements are established. To ensure that all teachers are able to fulfil the national and state-specific requirements in their everyday work, they are trained at universities over a period of five years in specialist subjects, didactics and vocational education. Aspiring nursing specialists are prepared by a curriculum developed by the federal state. As there is currently a shortage of teachers in the vocational subject area of nursing at vocational schools, the different federal states offer lateral entry and lateral entry programmes in order to attract teachers without the classic two-phase training (teacher training and preparatory service).

8.3.4   Non-regulated Profession: Nursing Teacher Trainer in Germany

Lecturers in the teacher training programme must have reflected on the professional challenges of university teaching and the case references (lecturers vs. teacher training students) and system references (lecturers vs. university/educational system/ social system) in addition to the two mentioned case references (nursing specialist vs. patient and teacher vs. trainee) and system references (nursing specialist and nursing setting, healthcare system, social system and teacher and vocational schools, educational system and social system). For these persons, the "double case reference" becomes a "triple case reference", which must be taken into account within teacher training (see Image 8.1).

Lecturers who train teachers in the first phase (scientific studies) are generally scientifically trained, rather than in higher education didactics. The personnel working at universities can be divided into two different status groups: scientific staff and professors. Scientific staff are not trained for teacher training. They did not have a single course during their studies on how aspiring teachers are trained "well". They have the specialised knowledge of the respective discipline, but at the beginning of their work in teacher training they do not have any formally acquired competences in the field of university didactics, how teacher trainees are trained, i.e. how seminars, exercises and internships should be designed so that the aspiring teachers are well prepared for their future tasks. They have also studied completely different study programmes, even though they all train teachers. Depending on whether the lecturers are employed in the field of specialist science or didactics of the vocational specialisation or the teaching subject or vocational pedagogy, they have previously dealt with pedagogical issues in different ways. Teachers in the field of nursing science have had little contact with the topics of education, training, teaching and learning in their own nursing science studies. An exception to this are the topics of patient counselling and patient education. But how teachers were supposed to be trained was not a question in the nursing science study programme. Employees who on the other hand offer classes in the field of vocational pedagogy or nursing pedagogy are familiar with the pedagogical content that they teach, but have also not dealt with the

Triple case reference

Double case reference

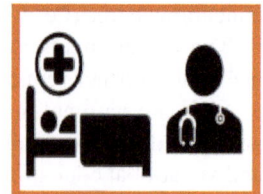
Single case reference

Image 8.1 Graphical representation of the meta-professional teaching and research concept (from Seltrecht & Thomas, 2023)

way in which this pedagogical content is taught at universities before starting their work in the field of nursing teacher training practice.

Unlike the status group of scientific staff, there is a regulation for the appointment of professors in the individual federal states that "only those who can prove three years of school practice or adequate pedagogical experience can be appointed as professors with educational science or specialised didactic tasks in teacher training. As an exception, it is also possible that experience in empirical research corresponding to the tasks may be recognised if proof of sufficient professional experience is provided within the first five years after the appointment. "(§ 35 Higher Education Act Saxony-Anhalt). Regardless of these required school experiences for university professors in teacher training, professors who have responsibility for nursing didactics are characterised by heterogeneous educational biographical, professional biographical and scientific-theoretical socialisation processes when compared to all German professors of nursing didactics. Some professors who currently develop and supervise study programmes for nursing teachers have completed their own nursing training, while others have not. Some have completed a nursing science programme, others a nursing pedagogics programme or a teaching programme, and others have socialised in the scientific theory of educational science. These different professional and scientific experiences lead to "the production of different mindsets, which ultimately find their way into the relevant teacher training and the nursing professions." (Ertl-Schmuck, 2018, p. 227).

8.4   Theoretical-Related Term of Competence and Practical-Related Competence Models

8.4.1   Theoretical-Related Term of Competence

Regardless of the regulation of a profession, the exercise of a profession is linked to the existence of profession-specific competences. The term "competence" has become increasingly important in the course of social developments and professional challenges. In occupational sociology, the individualisation of the working life has been a topic of discussion for several decades. In this context, education is understood as a life and career opportunity. It is even spoken of a „Arbeitskraftunternehmer" (entreployee) (Pongratz & Voß, 2003; Voß, 1998). The education policy demand for lifelong learning also makes the continuous training and further development of competence necessary. The term "competence" has become so ubiquitous in recent years that "being competent" is now considered a quality criteria even in everyday life.

The term competence was introduced into the scientific discussion by the linguist Noam Chomsky in 1973. In the context of the discussion about key qualifications (Mertens, 1974), it gained in importance. Today there is no universal understanding of competence, but the term is known throughout the various scientific disciplines—however, the meaning is different in each case. Therefore, Weinert (2003) also speaks of an "inflationary" term. Erpenbeck and Rosenstiel (2003) state that the term competence is "theory-related", i.e. it must always be redefined in the respective context of scientific work. In a scientific project, for example, reference must be made to concrete theories and concepts of an understanding of competence in order to emphasise in which sense the concept of competence is understood in the respective project work.

The term competence can, however, be compressed to common characteristics in the three related sciences of education, psychology and sociology: A subject's competence manifests itself in the performance of actions in which the subject defines the given situation and organises its own actions on the basis of this definition of the situation. Whether a person has competence is therefore only revealed when he or she masters situations appropriately and professionally. However, competence is not a static parameter—it is assumed to be capable of change and development. If this possibility for change were not inherent in competence, all pedagogical action would be in vain.

8.4.2 Practice-Related Concept of Competenc

The term 'competence' is not even clearly defined in practice-related models. This can be seen in the example of two reference frameworks: the European Qualifications Framework and the German Qualifications Framework. Both are reference systems for promoting lifelong learning and identifying different pathways—general education, adult education, vocational education and training and higher education—to achieve the different levels of competence. The EQF differentiates between knowledge, skills and competence:

- *Knowledge*: "In the context of EQF, knowledge is described as theoretical and/or factual." (EQF)
- *Skills*: "In the context of EQF, skills are described as cognitive (involving the use of logical, intuitive and creative thinking) and practical (involving manual dexterity and the use of methods, materials, tools and instruments)." (EQF)
- *Competence*: "In the context of EQF, competence is described in terms of responsibility and autonomy." (EQF)

The GQF distinguishes between professional competence, which consists of knowledge and skills, and personal competence, which consists of social competence and independence.

- In the area of *professional competence*, attention is focussed on the breadth and depth of knowledge acquired and the extent to which graduates possess skills. This refers to the ability to use and develop tools and methods. It also includes the ability to assess work results." (GQF)

- "*Personal competence* includes social aspects: Teamwork and leadership skills, the ability to contribute to shaping one's own learning or working environment, and communication skills. In addition, there is independence and responsibility, the ability to reflect and learning competence." (GQF)

The two reference systems alone represent different understandings of competence: On the European level, competence is categorised at the same level as knowledge and skills. On the German level, the term competence is an umbrella term that can be differentiated into various dimensions. In practical work, there are therefore various frames of reference to which we can refer. Similar to scientific works and the theory-based concept of competence used there, the concept of competence, which in this case is relative to practice, must also be defined here.

8.4.3 Practice-Related Competence Models in Nursing Practice

Competences relating to professional nursing are formulated in the Nursing Professions Act. Nursing training imparts the "professional and personal competences including the underlying methodological, social, intercultural, digital and communicative competences and the underlying learning competences as well as the ability to transfer knowledge and self-reflection. In this context, lifelong learning is understood as a process of one's own professional biography and continuous personal and professional development is recognised as necessary." (§ 5 PflBG).

Nursing scientist Patricia Benner has introduced a further systematisation into nursing science and nursing education by allocating competences to different competence levels: She distinguishes between the following five competence levels:

- Level 1: Novice,
- Level 2: Advanced Beginner,
- Level 3: Competent,
- Level 4: Proficient,
- Level 5: Expert (see Benner, 2017).

8.4.4 Practice-Related Competence Models in Nursing Education Practice

Selected teachers in nursing training in Germany are involved in the curricular work at the macro level, i.e. the development of framework curricula for the vocational training of nursing specialists at federal and state level. In Germany, these teachers are always involved in the further training of teachers with regard to the new curricula. In addition to this work on the macro level of curricular work, all teachers work at their school, i.e. on the meso level of curricular work, on the implementation of the national or state-wide curricula into the school's own curricula. They collectively determine which professional action situation (usually a specific professional action example/ case study) should be used for a learning situation and which specific competences should be imparted. Each teacher is then constantly involved in the planning of each individual lesson, i.e. at the micro-curricular level, by determining the competences and competence levels for each individual lesson, methodically implementing them and critically and constructively reflecting on them after the lesson.

However, a teacher at a vocational school does not only teach in nursing training. Instead, he or she is involved in other educational programmes. These other educational programmes have their own competency models, with which teachers must also be familiar. For example, a glance at the curricula of the federal state of Saxony-Anhalt for the vocational grammar school for health and social work, a secondary school at which a higher education entrance qualification can be obtained and which is relatively often subsequently followed by nursing training, already shows different competence models at the level of the individual teaching subject: For the subject health, for example, a distinction is made between analytical competence, evaluation competence and counselling competence. In the subject education/psychology, on the other hand, a distinction is made between analytical competence, evaluation competence and differentiation and decision-making competence. Teachers are therefore expected to juggle these competence models in the planning, implementation and reflection of their lessons.

8.4.5   Practice-Related Competence Models in Teacher Training Practice

Aspiring teachers in Germany are prepared for curricular work at a macro, meso and micro level during their teacher training programme: In addition to specialised academic training (e.g. nursing science) with 180 ECTS (equivalent to 5400 h), a teacher training course in Germany comprises 90 ECTS (= 2700 h) in vocational pedagogics and subject didactics in accordance with the requirements of the Conference of Ministers of Education and Cultural Affairs (KMK, 2017). Within this vocational pedagogical and didactic teacher training programme, students must complete three internships in which they deal with the professional tasks of teachers both theoretically and practically through work shadowing, discussions, reflection tasks, etc. They will also have their own initial teaching experiments for which lesson plans have to be written, i.e. curricular work has to be done at the micro level. Within the framework of two qualification theses (10 CP Bachelor's thesis, 20 CP Master's thesis), school- or teaching-related questions can be examined further theoretically or empirically. Following this first phase of teacher training (Bachelor's and Master's degrees), there is a second phase of up to 24 months, the traineeship (Referendariat).

With the aim of being able to compare teacher training programmes in the field of nursing at a national level and to promote convergence, the Conference of Ministers Education and Cultural Affairs, i.e. the association of education ministers of all federal states in Germany, presented a subject profile in 2015, which contains the subject-related competence profile on the one hand and study content from the fields of nursing science, health science, medicine, natural science and other related sciences (psychology, sociology, law, business studies) as well as study content from the field of nursing didactics on the other hand. As this subject profile of the Conference of Education Ministers is only binding for universities, but not for higher education institutions of applied sciences, where nursing teachers are also trained (unlike for other disciplines and teaching subjects), professors of nursing didactics have reached an agreement on the development of a recommendation framework for nursing didactics course components, which was developed in a three-year work process by 20 experts in nursing didactics and has been available since 2018 as the "Fachqualifikationsrahmen Pflegedidaktik" [FQR Pflegedidaktik (Specialised qualification framework nursing didactics)]. The FQR Nursing Didactics is regarded by those involved in the development process as a "reference framework for the conceptualisation of nursing didactic study components" (Dütthorn & Walter, 2018): "It should be expressly understood as a recommendation for the specific organisation of nursing didactic study components. This means that the FQR Nursing Didactics cannot be implemented identically and completely at every study location, with regard to individual module descriptions and learning outcomes. University or location-specific accentuations or prioritisation, summaries or extensions are both possible and desirable. It is about a common orientation, not a rigid standardisation of study programmes and contents of nursing didactics." (ibid.) This limitation in terms of range is also necessary because the FQR Nursing Didactics only focuses on the specific nursing didactic areas of study—but does not take into account other differently orientated areas of study (e.g. vocational pedagogics, second teaching subject/second vocational specialisation). From a professional theoretical perspective, the FQR Nursing Didactics fulfils two functions: It represents an instrument for the future collective professionalisation of nursing teacher training if it is used in the development of new study programmes and as an orientation framework for the accreditation of existing study programmes. At the same time, it is an expression of the already completed collective professionalisation of nursing teacher training, namely when representatives of almost all study locations in Germany where teachers for the nursing professions are trained agree on a consensus and set out which competences they consider to be important for the conception and further development of nursing didactic study components. The work of the 20 participating professors of nursing didactics shows that their work occurred on a macro-curricular level. Each individual professor is active at the meso level of curricular work at his or her location on the basis of this reference framework when developing the study and examination regulations and the corresponding module handbook. Each lecturer is in turn active at the micro level of curricular work on the basis of the study and examination regulations and the present module handbook when planning a course.

8.5   Discussion and Conclusion

The development of a training course in "Nursing with specialisation in geriatric nurs-ing" was relatively easy to implement at the college in Panjin, China, as, according to the college management, no legal conditions in the sense of a professional law had to be considered. The only thing that should be included were introductory compo-nents at the beginning of the programme in the sense of a country-specific Studium Generale. As such, the training course "Nursing with specialisation in geriatric nurs-ing" at Panjin College is a non-regulated profession. Also, experience at the college suggests that the teaching profession for the college is not regulated by the state. Regardless of whether a profession is regulated or not, each training programme team, whether at macro, meso or micro level, defines the learning outcomes, i.e. the competences that are to be developed with pedagogical intent, independently and autonomously. The members of the training programme team, i.e. the teachers, must therefore also have professional competence in the sense of curricula competence: Teachers—just as in other professions or in the professions they train for—must inde-pendently define the situation in the specific situation and then act accordingly. This competent acting is also necessary in the development of curricula—namely when the respective learning outcomes and associated learning content must be described for the respective target group and the associated competence developments must be determined both in relation to each other and according to time periods. As a conse-quence of these discussion results, professional competence with regard to curricular competence must be reflected at the macro level of a country, at the meso level of an educational institution and at the micro level of each teacher. As shown above, it is necessary for the development of curricula that teachers have professional compe-tence in order to achieve the competence enhancements for learners specified in the respective regulatory instruments, depending on the field of practice.

On the part of the Chinese college, there were no competence models to be consid-ered. On the one hand, this made it easy for teachers, because they were free to create new competence models. On the other hand, this freedom also made it more difficult to work, as the possibility of creating one's own curriculum free from existing models also meant taking on the curricular work independently and on one's own responsi-bility in order to eventually have a curriculum according to which the next generations of nursing specialists would be trained at the college in the future. The fact that most of the teachers did not have any training in nursing didactics, however, meant that the Chinese teachers first had to undergo an extensive introduction to the subject of didactics and curriculum development, including an examination of the concept of competence and various competence models. In the beginning, the Chinese educa-tion programme team wished to be oriented towards the German standard. After an advanced training course had been held to communicate the difference between the statutory professional law regulated at federal level and the associated training and examination regulations for the (then still three) nursing professions on the one hand and curricular implementation at federal state level due to the existing federalism in Germany as well as a further design of the state curriculum at school level, the

Chinese training programme team at Panjin College independently developed their curriculum.

The college in China was very interested in the consideration of both case and system references in the training of aspiring nursing specialists. Work-related situations from the context of nursing activities were constructed and integrated into the newly designed training programme as part of the new curriculum. For this purpose, workshops were held with the teachers so that teachers with nursing experience had the opportunity to contribute case presentations, which were subsequently didactically prepared. As there was little or no knowledge of the field of geriatric care in terms of the care setting (acute inpatient geriatrics, nursing home) and the requirements of caring for elderly and very elderly, mostly multimorbid patients, excursions were undertaken, e.g. to a geriatric ward and a nursing home.

The experiences gained from the project indicate that teachers at the Chinese college have not yet been trained to the same breadth and depth in terms of vocational education and didactics as teachers for vocational schools in Germany. Furthermore, teachers have hardly ever had the experience of developing curricula on their own responsibility or on behalf of other teachers—by being commissioned to do so by the responsible ministry or the respective school—which they themselves are responsible for, representing these to representatives from politics, science and professional practice and, if necessary, communicating them to other teachers through further training events. These differences, resulting from the project experiences, will ultimately show that the development of teachers' professional competence in curriculum development must not begin with further teacher training, but with the initial teacher training programme. The requirements for Chinese teachers therefore differ from those of teachers trained in Germany. In Germany, the professional understanding of teachers is already addressed during their studies and they are trained to act independently towards the target group of learners. The "balancing act" described in the professional theory between expectations at system level (institution, society) on the one hand and demands on pedagogical acting towards the target group of learners (case reference) must be constantly explored and reflected upon anew. In Germany, the learning needs of the trainees and the idea of teaching the trainees all the professional competences that they need to work independently and responsibly in a self-determined way with a selection of tools and methods for which the teacher is responsible are the guiding principles. This article shows that the curricular work of teachers is very complex and therefore requires extensive training, particularly in vocational pedagogics and specialised didactics. In Germany, teacher training in the vocational specialisation of nursing has two curricula, the subject profile presented by the Conference of Education Ministers and the subject qualification framework for nursing didactics developed by a group of 20 experts, which provide orientation for teacher training in Germany. Teachers who have been trained in this manner have acquired the necessary competences to juggle with the various competence models that are repeatedly presented in the specialist literature and to develop or use competence models themselves in a "practice-related" manner for other target groups (e.g. trainees in the nursing professions) in order to design and accompany teaching–learning processes in a professional and appropriate manner.

The reflection on the relationship between system reference and case reference and the formation of a professional identity as a member of the profession who is aware of his or her task (mandate) and permission (licence) for pedagogical acting and continuously works on this is necessary in order to independently define competences in curricula. These competences, however, cannot be changed in further training courses, and certainly not those coming from abroad. In the context of an educational export from Germany to China, it is only possible to gain insights into the self-image of German teachers, their core competences and the logic of their actions. The introduction of a "curriculum according to German standards" then also requires more than just the presentation of a paper with competence descriptions and the naming of knowledge components that are to be taught. What constitutes the "professional competence" of Chinese teachers must be answered by the Chinese side first, and only in the context of the relationship between society, the (teaching and nursing) profession and (teaching and nursing) education. Against this background, considerations regarding the didactics of higher education should also be made as to how a curriculum for teacher training should be realised. At this point, the course is also set for to what extent teachers should be involved independently and autonomously in curricular work at macro, meso and micro level. It is therefore important to initiate a discussion on teacher training so that these teachers can responsibly develop and implement the curricula for vocational education and training.

With regard to Germany, however, this article also self-critically demonstrates that scientific staff at universities who train teachers (in distinction to professors) have not received any training in higher education didactics when they start training aspiring teachers. A discussion on professional teacher training and the professionalisation of teacher training staff is therefore urgently needed. Thanks to the international exchange between China and Germany, this gap in the triple case and system reference became clearly evident on the German side.

Literature

Benner, P. (2017). *Stufen zur Pflegekompetenz: From novice to expert*. Hogrefe AG.

Dütthorn, N., & Walter, A. (Hrsg.) (2018). *Fachqualifikationsrahmen Pflegedidaktik. Deutsche Gesellschaft für Pflegewissenschaft*. Online-Ressource.

Erpenbeck, J., & Rosenstiel, L. v. (2003) (Ed.), *Handbuch Kompetenzmessung*. Stuttgart Schäffer-Poeschel.

Ertl-Schmuck, R. (2018). Medizinpädagogik – ein diffuser und obsoleter Begriff im Wandel der Zeit. In H. Ohlbrecht & A. Seltrecht (Eds.), *Medizinische Soziologie trifft Medizinische Pädagogik* (pp. 215–229). Springer.

KMK—Kultusministerkonferenz. (2017). *Ländergemeinsame inhaltliche Anforderungen für die Fachwissenschaften und Fachdidaktiken in der Lehrerbildung*. https://www.kmk.org/fileadmin/Dateien/veroeffentlichungen_beschluesse/2008/2008_10_16-Fachprofile-Lehrerbildung.pdf. Abfrage: February 13, 2018

Mertens, D. (1974). Schlüsselqualifikationen. Thesen zur Schulung für eine moderne Gesellschaft. *Mitteilungen Aus Arbeitsmarkt- und Berufsforschung. H., 1*, 36–43.

Pongratz, H. J., & Voß, G. G. (2003). From employee to 'entreployee'. Towards a 'self-entrepreneurial' work force? In *Concepts and Transformation* (Vol. 8:3, pp. 239–254). John Benjamins Publishing Company.

Seltrecht, A. (2015). Der „doppelte Fallbezug" – Herausforderung in der Lehramtsausbildung in der beruflichen Fachrichtung Gesundheit und Pflege. In: Jenewein, Klaus/Henning, Herbert (Hrsg.), *Kompetenzorientierte Lehrerbildung. Neue Handlungsansätze für die Lernorte im Lehramt an berufsbildenden Schulen* (pp. S. 209–227). W. Bertelsmann Verlag.

Seltrecht, A. (2024). Berufliche Didaktik personenbezogener Berufe. In: Spöttl, G, & Tärre, M. (Hrsg.), *Didaktiken der beruflichen Aus- und Weiterbildung* (pp. 623–632). Springer.

Seltrecht, A., & Thomas, V. (2023). *Einfluss von Weiterbildung auf die individuelle und kollektive Professionalisierung und De-Professionalisierung von Berufen (EWik)*. Poster.

Voß, G. (1998). Die Entgrenzung von Arbeit und Arbeitskraft. Eine subjektorientierte Interpretation des Wandels der Arbeit. *Mitteilungen Für Arbeitsmarkt- und Berufsforschung, H., 4*, 501–522.

Weinert, F. E. (2003). Concept of competences: a conceptual clarification. In: Rychen, D.S., & Salgnik, L.H. (Ed.), *Key competencies* (pp. 45–65). Hografe & Huber.

Astrid Seltrecht degree in educational sciences, studied educational sciences, sociology and psychology at the Free University of Berlin, received her doctorate at the Goethe University Frankfurt/Main and has been a university professor for didactics in health and nursing sciences at the Otto von Guericke University Magdeburg, Germany, since 2018. She is responsible for vocational schoolteachers training in the fields of health and care in the master's degree. In the PhD Programme "Vocational training and personnel development" she is responsible for the "Professionalisation of and professionalism in personal relationships". Her research focuses on didactic research on school and teaching as well as the educational analysis of occupational safety, health protection and accident protection.

Chapter 9
Coordinated Development of the Beijing-Tianjin-Hebei Region—Research into How the Elderly Care Industry and Vocational Education in the Field of Gerontology Are Being Integrated—A Case Study of Tianjin City Vocational College

Jian Fu

9.1 Research Background

Showing a common concern about the elderly care industry, the Ministry of Education, the Ministry of Civil Affairs, the National Development and Reform Commission, the Ministry of Finance, the Ministry of Human Resources and Social Security, the National Health and Family Planning Commission, the Central Civilisation Office, the Central Guidance Commission on Building Spiritual Civilisation, the Communist Youth League, and the Central National Office for Ageing jointly released their "Opinions on Accelerating the Development of Talents for the Elderly Care Industry" in July 2014. The communique encouraged the vigorous development of courses relating to elderly care, as well as the continued expansion of human resources training, the strengthening of professional facilities related to the industry, the acceleration of the establishment of an elderly care human resources training system, and extensive improvement to the quality of human resources training in the industry. It also called for all actors to adapt to the developing needs of the elderly care industry—all the while following the principals of "positive development, extensive enhancement, using multiple models, and a focussed approach".

In April 2015, the Central Politburo of the Communist Party of China reviewed and approved its "Outline of the Plan for Coordinated Development for the Beijing-Tianjin-Hebei Region (hereinafter the BJ-TJ-HE)". The Outline made coordinated

J. Fu (✉)
Tianjin City Vocational College, Tianjin, China
e-mail: 34257822@qq.com

© The Author(s) 2025
Deutsche Gesellschaft für Internationale Zusammenarbeit (GIZ) GmbH et al., (eds.),
Sustainable Aging, https://doi.org/10.1007/978-3-662-69139-7_9

development for the region a top national strategy. "The number of elderly people over 60 years old in Beijing, Tianjin and Hebei has exceeded 16.3 million. Among them, Beijing is home to 3 million elderly people and Tianjin 2.15 million. The elderly care industry is under tremendous pressure. The BJ-TJ-HE region must work together to coordinate the development of the elderly care industry."

In this context of integrating resources from multiple entities, offering vocational education for the elderly care industry, and bringing about the innovative development of vocational schools in the three regions; two key aspects remain vital:

1. an improvement of the quality of training of elderly care professionals;
2. and the promotion of the integration of vocational education and the elderly care industry to ensure the coordinated development of vocational education in the region.

9.2   The Research Process

Tianjin City Vocational College serves as National Vocational Education Teaching Guidance Committee Member of the Elderly Service and Management Professional Teaching Guidance Committee Bureau, and with support from the Department of Vocational and Adult Education of the Ministry of Education and the National Industry Guidance Committee, as well as guidance from the Tianjin Education Commission, it took the lead in organising and establishing in 2015 the "Training for Elderly Care Professionals and Industry-Education Collaborative Committee for the BJ-TJ-HE Region" (hereinafter the Industry-Education Collaborative Committee). The Committee's collaborative efforts help to translate industry needs into educational offerings in the field of gerontology and training of elderly care professionals as part of the coordinated development of the BJ-TJ-HE region, which has vastly improved the level of the College's gerontology services and management course programme. During the construction of the 13th Five-Year Plan's "Double-First Class" in Tianjin, the College was approved as a world-class, domestic first-class higher vocational college. Its course programme in "Gerontology Services and Management" was classed as an exemplary and top-quality course in Tianjin.

The College has held three consecutive fora on the topic of "Building a Modern Vocational Education and Modern Service Industry for the Coordinated Development of the BJ-TJ-HE Region—Integration of Industry into Education", and followed these with the organisation and establishment of the "Tianjin Elderly Care Service Industry Vocational Education Teaching Steering Committee" (hereinafter the Tianjin Elderly Industry Steering Committee), the "Tianjin Elderly Education Guidance Center" and the "BJ-TJ-HE Health Industry Technology Collaborative Innovation Center".

Such entities as the Industry-Education Collaborative Committee and events like the fora on industry-education integration, have led to the establishment of the so-called "five parties and five industries" collaboration. The five parties—government, industry, businesses, vocational schools and research institutes—form a cooperation mechanism; and the five industries—production, industry, enterprise, occupation and profession—form an operational mechanism. This has created a multi-party collaboration, with the integration of industry into education, and a fusion of learning and practical training, allowing Beijing, Tianjin and Hebei to play a leading role in the formulation of elderly care industry policy. Progress has been made from the initial establishment of coordinated institutions, which has created a cooperation mechanism of "five parties and five industries working together"; to the integrated synergy of colleges and universities and the establishment of a quality improvement platform for elderly care professionals; through to the integration of the civil affairs coordination system, the integration of industry into education; the uniting of research institutes in the BJ-TJ-HE region, to developing channels for international exchange, and promoting the in-depth development of gerontology course programmes and innovative and practical training in all aspects, taking into account multiple perspectives, and maintaining high quality.

9.3 Main Research Problems

The following key problems need to be resolved when developing gerontology course programmes and building innovative, practical training projects for the BJ-TJ-HE region:

1. Issues concerning the coordinated integration of industry and education in the training of elderly care professionals for the BJ-TJ-HE Region;
2. Issues relating to the sharing of course resources and the creation of a competition platform to boost skills between professional vocational colleges in the BJ-TJ-HE Region;
3. Issues concerning innovation and scientific research cooperation in the coordinated training of elderly care professionals for the BJ-TJ-HE Region;
4. Issues relating to the implementation of a modern apprenticeship pilot programme launched by the Ministry of Education and the introduction of social services offered by teachers and students for Beijing-Tianjin-Hebei Region;
5. Issues concerning the reliance on collaborative resources of the BJ-TJ-HE region to introduce professional internationalised courses as well as promoting international exchanges and improving the level of teaching.

9.4   Research Project Results

9.4.1   Establishment of a Collaborative Organisation in the BJ-TJ-HE Region, Creating a Cooperation Mechanism Formed by the So-Called "Five Parties and Five Industries"

Establishment of a Coordination Committee for the Coordinated Integration of Industry and Education in the Training of Elderly Care Professionals in the BJ-TJ-HE Region, to Promote the Integration of the Elderly Care Industry and Education

The "Elderly Care Professionals Training and Industry-Education Collaborative Committee for the BJ-TJ-HE Region" was set up in May 2015, with support from the Department of Vocational and Adult Education of the Ministry of Education and the National Industry Steering Committee. Its members consist of 47 colleges, enterprises and social organisations from the region. The Industry-Education Collaborative Committee has held three consecutive fora on the topic of "Building a Modern Vocational Education and Modern Service Industry for the Coordinated Development of the BJ-TJ-HE Region—Integration of Industry into Education", with Tongfang Knowledge Network Technology Co., Ltd building a network collaboration platform. The platform has helped guide the three regions to establish a cooperation and operational mechanism—a collaboration of the so-called "five parties and five industries." To properly integrate resources and provide a platform for exchange for the elderly care service industry, the educational resources of colleges in the region are linked with outstanding enterprises from the elderly care industry nationwide, connecting industry with education and allowing for collaboration between schools and enterprises. This ensures resource sharing and complementarity between vocational colleges and elderly care enterprises in the BJ-TJ-HE region in the fields of elderly care talent, knowledge, technology and equipment, which in turn accelerates the integration of industry into education—bringing modern vocational education in line with the needs of the elderly care service industry.

Establishment of the Tianjin Industry Steering Committee to Strengthen Guidance of the Elderly Care Industry

In February 2017, the Bureau of Civil Affairs of Tianjin and the Tianjin Municipal Education Commission established the Tianjin Elderly Care Industry Vocational Education Teaching Steering Committee (hereafter the Tianjin Steering Committee) designed to research, guide, service and consult for vocational education and teaching in the elderly care industry in Tianjin, as well as to steer the vocational education and training work of the entire city's elderly care service industry. The Committee comprises 22 expert members from undergraduate vocational colleges and industry association entrepreneurs to experts from scientific research institutions and senior university representatives. The Tianjin Steering Committee has established and improved the government-led, industry-led, and enterprise-participating vocational education mechanism through measures such as joint committee meetings, and has further strengthened the guiding function of the industry in the development of a modern vocational education system, and this during the reform and development of vocational education. The Committee has also played an active and effective role in promoting the development of city-level geriatric education and elderly care resource pools, as well as improving the quality of training for elderly care professionals.

Establishment of the Tianjin Elderly Education Guidance Center, to Promote the Elderly Care Profession

In February 2017, relying on the elderly care coordination platform in the BJ-TJ-HE region, the Tianjin Geriatric Education Guidance Centre was established by Tianjin City Vocational College's first vocational education group to promote regional community education and service lifelong learning. The establishment of this Centre will give full play to the advantages of vocational education resources in the downtown area of Tianjin, and will strengthen the development of the gerontology curriculum, reinforce the development of the geriatric education programme team, and actively promote theoretical research and the application of geriatric education in the workplace. It will further promote vocational education and continuing education in the BJ-TJ-HE Region, as well as the coordinated development and deep integration of geriatric education, providing support and services for gerontology and the elderly care industry in the region with a large backing from education.

9.4.2 Achieving Synergy Among Vocational Colleges in the BJ-TJ-HE Region to Establish a Platform to Improve the Quality of Gerontology Course Programmes

Creation of a Competition Platform in the BJ-TJ-HE Region to Promote Elderly Care Teaching Through Competition

With support from the Industry-Education Collaborative Committee, and having effectively established synergies between professional colleges and universities offering gerontology courses in the BJ-TJ-HE region, a competition platform for nursing care skills was created—the 2016 Seventh National Vocational Colleges Civil Affairs Vocational Skills Competition for the Elderly Caregiver. The competition has various aims: firstly, that it will lead the way in setting standards and will become a learning channel, so that all colleges and universities can understand and communicate with each other through competition; secondly that it will help identify gaps through comparison and provide a comprehensive assessment of the teaching achievements of participating colleges, so the latter may perceive their own strengths and weaknesses; and finally to promote teaching through competition and further optimise and adjust the methods of running a school, human resources training programmes, teaching content, teaching methods, and ultimately to shape modern, innovative elderly care professionals who are suitable, valuable, practical and useful.

Promoting Co-education Between Schools and Enterprises Through a Modern Apprenticeship Pilot Platform

Further demonstrating the effective role of the Industry-Education Collaborative Committee, is the creation of the "Modern Apprenticeship Practice Teaching Center", which consists of 11 elderly care institutions, enterprises supporting elderly care facilities, elderly care service platform enterprises and home-based residential care services in the BJ-TJ-HE region. On top of this initiative, a second batch of modern apprenticeship pilot projects of the Ministry of Education was carried out by members of the BJ-TJ-HE Collaboration Committee including Tianjin Tangbang Technology Co., Ltd., North Keye Live House Service Promotion Center, and Tianjin Leyue Smart Care Co., Ltd. A "modern apprenticeship" cooperation agreement was signed to achieve integration between schools and enterprises, resource sharing, tutor training and apprentice learning in both schools and enterprises and integrating the production process and teaching process. It allows for joint discussion in the formulation and revision of course training programmes, for joint implementation of curriculum reforms, expansion and development, and lays a solid foundation for the joint training of elderly care professionals under the modern apprenticeship model.

Sharing Quality Education Resources Through an Information Resource Library

The Chair of the Industry-Education Collaborative Committee—Beijing Social Management Vocational College—and Beijing Labour and Social Security Vocational College have jointly developed the "Elderly Services and Management Course Resource Library Project for the Ministry of Education", and have presided over the development of the "Operational Management of Elderly Care Institutions" project as well as participating in the construction of a further two projects, namely "Social Work for the Elderly" and "Laws, Regulations and Standards for the Elderly". They are responsible for the "Tianjin Elderly Care Services and Elderly Education Resource Sharing Development Project", with a total investment of RMB 6.2 million (EUR 870,000), which includes multi-module course resources relating to course programme, curriculum teaching, virtual simulation, practical teaching, and industry and enterprise resources. The institutions in the BJ-TJ-HE region have effectively established a course resource sharing mechanism for interoperability, mutual benefit, and mutual learning. Furthermore, they can study gerontology course programme resources, core curriculum resources and training enhancement resources via an online platform known as "smart vocational education" which allows for interactive resource sharing among academic institutions.

9.4.3 Integrating the Coordination System Between the Civil Administrations in the BJ-TJ-HE Region to Link Industry with Elderly Care Education

Working Together with the Civil Administrations of the BJ-TJ-HE Region to Deepen the Practice of Industry and Education Integration

The College actively participated in the "Joint Meetings of Collaborative Development of Elderly Care Work in the BJ-TJ-HE Region", organised by the civil administration system, regularly sharing typical experiences and exchanges. The "Joint Meetings" initiated by the Civil Affairs Department (Bureau) of the BJ-TJ-HE Region aim to promote communication between elderly care enterprises, as well as between schools and enterprises in the three regions, to strengthen the management of elderly care institutions, enhance the training of elderly care professionals, and improve the quality of elderly care services. The platform to integrate schools and enterprises within the civil administration system, has effectively expanded communication channels and the space for cooperation between the College and enterprises engaged in the elderly care industry in the BJ-TJ-HE region. The platform has enabled the collection and analysis of comprehensive data from over one thousand elderly care institutions of various types, as well as the evaluation of local elderly care policies in the different regions. Furthermore, it has, together with some of the elderly care service enterprises, reached a cooperation agreement for modern elderly care personnel training, successfully improving the effectiveness of the integration between schools and enterprises and deepening the integration of industry into education.

Combining Industry with Education and Focusing on Work-Integrated Learning, to Develop Social Services for Teachers and Students

Consistent with what the Industry-Education Collaborative Committee promotes, the College has introduced the concept of vocational training throughout its work to integrate schools and enterprises as well as in the field of social services. It focusses on work-integrated learning and combining study with training, for collaborative innovation. The College has signed off-campus training centre agreements with Beijing Chenghe Jingchang Residence, Beijing Taikang Home (Yanyuan) and other enterprises, to allow students to alternate between study and work, and provide teachers the opportunity to provide social services. Taking advantage of the resources of a high-quality gerontology teaching team and advanced equipment within the elderly care training centre, the collaboration between schools and enterprises provides extensive training opportunities for elderly care nursing staff, social workers, Chinese medicine rehabilitation physiotherapists and Tianjin Women's Federation, with more than 1,500 people trained on average each year. A service-oriented public welfare programme for the elderly was launched in cooperation with Jiangdu Road Residential District—"Health, Care and Fitness" home-based residential care services and training. Classes specialising in gerontology, working in tandem with the Heng-shanli and Leshanli neighbourhood committees in the Hebei region, were offered, and provided a variety of welfare services for the elderly in the community, from computer and Internet training, to cultural and sports activities.

Targeting the Real Needs of Wei County to Accurately Implement the Government's Poverty Alleviation programme

To implement the spirit of the Ministry of Education's "BJ-TJ-HE support for the implementation of vocational education and continuing education in Qinglong County and Wei County, Hebei Province (2018–2020)" and under the guidance of the Department of Vocational and Adult Education of the Ministry of Education, the College carried out a poverty alleviation programme in collaboration with Wei County, Hebei Province. The College effectively linked up with the Wei County Civil Affairs and local Education Bureau to carry out a full investigation of the Wei County Vocational Education Centre and the First Nursing Home of Wei County Civil Affairs Bureau. Following discussions on poverty alleviation projects, it was agreed to dispatch outstanding industry experts to Wei County to conduct professional training for local personnel engaged in providing security, cleaning, and nannying services, as well as elderly care nursing staff, and for some trainers to visit large-scale elderly care institutions in Tianjin. The results have enhanced the vision and professional practical experience of trainers, improved the overall quality of security, cleaning and nannying personnel, and enhanced the educational capacity of schools in the county, as well as expanding employment channels, and completing the task of poverty alleviation of the local people.

9.4.4 Uniting Research Institutes in the BJ-TJ-HE Region to Develop an International Exchange Channel for Elderly Care

Establishment of the BJ-TJ-HE Collaborative Innovation Centre for Healthcare Industrial Technology, to Develop Cooperation in Scientific Research Projects

The Vocational College, together with the Beijing Institute of Science and Technology, Hebei University of Technology, and Hebei University of Economics and Business have collectively formed the "BJ-TJ-HE Collaborative Innovation Centre for Healthcare Industrial Technology", which cultivates innovative young talent in elderly care, undertakes government and market research development projects, and carries out international exchanges and cooperation, to create a brand of BJ-TJ-HE elderly care service products and resource sharing. The College's gerontology programme collaborates with partners to carry out cross-cutting research projects such as the "BJ-TJ-HE Joint Training of Elderly Care Professionals", and the EU joint project for "Professional Talent Development Training in the Elderly Health-care Field and Curriculum Development for Student Training". It is also collaborating on a set of textbooks for a fitness series (sports and acupressure) in Taiwan and is applying for one related patent, laying a solid base for further international exchanges and cooperation.

Joint Creation of a Teachers Exchange Platform to Expand International Horizons

A teacher exchange platform has been set up in cooperation with Beijing Science and Technology Research Institute. A team of professional teachers specialising in gerontology participated in the 3rd International Forum on Ageing Service Science and Innovation and the 2017 BRICS Ageing Conference hosted by Beijing Yiyang Technology Co., Ltd.—a company affiliated to the Beijing Science and Technology Research Institute. Their participation effectively enhanced the international vision of the team of elderly care professional teachers. At the same time, teams of elderly care experts from Australia, Finland, Spain and Taiwan came on exchange visits to the College, and jointly organised the 2017 Tianjin-Taiwan Elderly Care Collaboration Forum with the College, discussing possible cooperation areas such as introducing an international curriculum, exchange visits and personnel training, thus expanding the development space for the internationalisation of the elderly care profession.

Introduction of Taiwanese and International Courses to Optimise the Residential Care Curriculum System

To cultivate the core skills of students specialising in home-based residential care services and enhance the internationalisation of teachers and students, the College collaborated with Beijing Yiyang Technology Co., Ltd. to introduce the EU International Gerontology Course (including international nursing, patient-centered care concepts, advanced skills and health assessment modules), Taiwan Physical Fitness programme (also known as exercise therapy) home-based residential care curriculum (including physical fitness, music therapy, cognitive training, horticultural therapy, reminiscence therapy, brain health courses) and Taiwan Dementia Management Coaching Course. Through these various types of training courses, teachers and students have improved their systematic understanding of international elderly care and have enriched their professional skills to deal with elderly groups requiring different home care needs. A total of 60 teachers and students have obtained the Taiwan Dementia Management Coaching Certificate and the EU-certified International Course Certificate for Geriatric Care, which has effectively optimised the home-based residential care course system for this major.

9.5   Project Research Innovation Points

9.5.1   Development of a Community of Elderly Care Professionals in the BJ-TJ-HE Region

With the establishment of the Industry-Education Collaborative Committee, the Tianjin Industry Guidance Committee, and the Tianjin Geriatric Education Guidance Center, as well as the three consecutively organised fora, member organisations have effectively improved their understanding of China's national strategy, and have shared new concepts, information and recruiting methods relating to elderly care vocational education in Beijing, Tianjin and Hebei. This has enhanced the sense of responsibility of professional caregivers, enforced their mission and sense of urgency, and allowed for joint research, joint development, sharing of ideas and experiences, joint use of resources and a win–win for the community of elderly care professionals in the BJ-TJ-HE region.

9.5.2   Joint Education Mechanism for Elderly Care Professionals in the BJ-TJ-HE Region—"Linking Five Parties and Five Industries"

Members of the "Elderly Care Vocational Training and Industry-Education Collaborative Committee in the BJ-TJ-HE region" are focussed on the integration of the elderly care industry into education. Following the so-called collaboration of "five parties and five industries", which form a cooperation mechanism and an operational mechanism respectively, there has gradually been a coordinated, innovative development in the integration model between industry and education. These concerted efforts have led to multiple schools and colleges in Beijing, Tianjin and Hebei producing graduates majoring in the same field, multiple enterprises offering a range of professional positions, and the development of talent across the three regions.

9.5.3   Collaborative Elderly Care Innovation and Service Platform in the BJ-TJ-HE Region

Effectively joining the "Healthcare Industry Technology Coordination and Innovation Centre for the BJ-TJ-HE region" led by the Beijing Science and Technology Research Institute and the "Joint Meeting of Collaborative Development of Elderly Care Work in the BJ-TJ-HE Region" led by the Civil Administrations of the three regions, an open platform and ecosystem has been established for collaborative innovation between government, industry, enterprises, vocational schools and research institutes in the region. Moreover, by building trust and highlighting the benefit of sharing, outstanding resources from the BJ-TJ-HE region have been brought together, to integrate fragmented industrial chains and innovation systems to form endogenous and sustainable innovation models and mechanisms which accelerate the innovation process of the aged service industry. This has advanced the innovation of relevant organisations, systems and policies, and boosted the development of the social ageing service system.

9.5.4 Paving the Way for Vocational Training of Elderly Care Service Personnel Within the Collaborative Development of the BJ-TJ-HE Region

Capitalising on the mechanism of the "five parties and five industries", the development of the elderly care coordination project in the BJ-TJ-HE region, together with the work to build the elderly care curriculum and share resources between the three regions, has effectively brought about work-integrated learning, and collaboration between schools and enterprises. This has allowed for resource sharing and advantageous complementarity between the region's vocational colleges and elderly care enterprises in the fields of human resource talent, knowledge, technology and curriculum, giving full play to the advantages of working in a group, to the effects of combined and scaled efforts, and paving the way for coordinated development in talent training innovation.

9.6 Practical Application of Research Findings

9.6.1 Applying the So-Called "Five Parties and Five Industries" Mechanism to the Activities of the Collaborative Institutions

Under the leadership of the BJ-TJ-HE Elderly Care Vocational Training and Industry-Education Integration Collaborative Committee, a further three fora were held under the banner "Building a Modern Vocational Education and Modern Service Industry for the Coordinated Development of the BJ-TJ-HE Region—Integration of Industry into Education". Leaders from relevant ministries participated, including from the education, civil affairs, human resources and social security, health and family planning ministries, and the women's federation, as well as well-known enterprises engaged in the elderly care service industry in the BJ-TJ-HE region and representatives of undergraduate, higher vocational and secondary vocational schools offering relevant majors. Almost 150 people participated in the discussions, successfully accomplishing joint research, joint development, sharing of ideas and experiences, joint use of resources and a win–win for the training of elderly care professionals in the BJ-TJ-HE region and the elderly care industry.

Under the leadership of the Tianjin Elderly Care Industry Steering Committee, the "Tianjin Elderly Care and Elderly Education Resource Sharing Conference" was held to attract and gather both key players in society as well as quality resources to integrate into vocational education and continuing education through the collaboration of "Government, industry, enterprises, vocational schools and research institutes", jointly serving elderly education and the elderly care sector.

Under the leadership of the Tianjin Elderly Education Guidance Center, the "2017 BJ-TJ-HE Elderly Education Promotion Seminar and Elderly Education Report Fair" was held, following the concept of collaboration in vocational education and continuing education, and creating an effective platform to share experiences and learning practices on the coordinated development of strategy in the areas of vocational education, elderly education and community education in the BJ-TJ-HE region.

9.6.2 Maximising Quality to Enhance the Impact of the "Three Platforms" and Increase the Effectiveness of Vocational Education

Taking advantage of the Tianjin-Hebei elderly care competition platform, the "7th National Vocational College Civil Affairs Vocational Skills Competition for Elderly Care Nursing Staff" was held, with the participation of 359 candidates from 65 vocational colleges nationwide. Students from the College's gerontology course have, for the past three consecutive years, successfully won one first prise, five second prises, and four third prises both in the national Nursing Care Professional Skills Competition and in the competition at the BJ-TJ-HE regional level.

Taking advantage of the leading role of the modern apprenticeship pilot programme of the Ministry of Education, the College signed a "modern apprenticeship" cooperation agreement with five members of the BJ-TJ-HE Collaboration Committee to explore a modern apprenticeship training model. Eight general managers and business leaders have been employed under the scheme, making over 20 visits to the College to teach the core curriculum as part of the integration between schools and enterprises, which has given a boost to the vocational training of elderly care professionals.

Following the College's participation in the development of three courses, including the "Operational Management of Elderly Care Institutions" in the Ministry of Education's Elderly Care Resource Library, it has promoted the development of a further 10 core subjects, to enhance the role of the information resource sharing platform, including an "Elderly Activity Organisation and Planning" course in the Tianjin Elderly Care Resource Library. Furthermore, schools and enterprises have jointly developed a set of school-based course resources with entities such as Tang-bang Science and Technology, and Beijing LoHo House Geriatric Care Service Promotion Center, focused on the core competence of promoting home-centered and community-based elderly care.

9.6.3 Capitalising on the Coordinated Development of Civil Affairs in Beijing, Tianjin and Hebei to Promote the Integration of the Elderly Care Industry into Education

Fully integrated into the civil affairs coordination system for the BJ-TJ-HE region, the College became a member of the National Vocational Education Teaching Guidance Committee of the Elderly Service and Management Professional Teaching Guidance Committee, and participated in two joint conferences on "Elderly Care Work for the Coordinated Development of the BJ-TJ-HE region" as well as three editions of the "National Elderly Care Industry and Vocational Education High Level Dialogue Activities". It has made recommendations for the development of the course curriculum, and expanded the influence of the course programme, which has now become a focal point for many related enterprises and institutions.

The College combines teachers' practical experiences in elderly care services and rehabilitation institutions with social services, and develops training resource packages encompassing the theoretical knowledge of the elderly care workers, practical assessments and professional competency-based surveys, while collaborating with neighbourhoods, communities and nursing homes to develop the so-called "three skills" social training services—technology development, technical consultation and technical services—which have already achieved a certain degree of success.

Linking industry to education has effectively improved the professional ability of the faculty team. The faculty team upgrades curriculum on various core programmes such as "Social Work for the Elderly—Methods and Practices" and develops learning resources such as training programmes and micro-courses. A textbook has been published for the 12th Five-Year Plan entitled "Laws, Regulations and Standards for the Elderly", a further coursebook is being edited for the 12th Five-Year Plan entitled "Administration and Management Practices of Elderly Care Institutions", as well as a coursebook entitled "Social Work for the Elderly".

9.6.4 Capitalising on the Collaboration of Research Institutes in the BJ-TJ-HE Region to Make Course Programmes More International

Course teachers participated in the "3rd International Forum on Senior Care Science and Innovation" organised by the Beijing Science and Technology Research Institute as well as the "2nd BJ-TJ-HE Elderly Care Professionals Coordinated Development Summit" and the "Sino-Japanese Advanced Seminar on Senior Care Technology for Dementia Sufferers" organised by the Training Centre of the Ministry of Civil Affairs among many other high-profile conferences and training sessions in Beijing, Tianjin and Hebei. Such participation has brought an international perspective to concepts, methods and content applied to the curriculum and has expanded their professional international vision.

Huang Yaoting, a clinical psychologist at the Changhua Christian Hospital in Taiwan, and Chunman Chen, Secretary General of the Changhua County "Solidarity with Dementia Association", have been enlisted to teach the "Alzheimer's Management Coaching Course". Nick Guldemond, Project Coordinator for the EU programme for Active and Healthy Ageing was also enlisted to teach courses on international geriatric nursing. Over 60 teachers and students have obtained corresponding course certificates, enriching the international elements of the course.

Thanks to the introduction of international courses and international cooperation and exchanges, the practical ability of gerontology course teachers and students has been significantly enhanced. Thirty-three fresh graduates were recruited by Taikang Life Insurance and Vanke Group's exclusive elderly care institutions and Beijing Chenghe Jingchang Residence among other core state-owned enterprises managing Beijing's elderly care service sites. The teaching team is also engaged in the development of elderly care professional skills training and community services in the workplace. The services reach an average of 1500 people per year and have won unanimous praise from society and partnering enterprises.

9.7   Conclusion

In conclusion, on the fourth anniversary of the keynote speech made by General Secretary Jinping Xi on the coordinated development of the BJ-TJ-HE Region, there have been discussions on what tasks are required and how to realise integration of the elderly care service industry into modern vocational education in the BJ-TJ-HE region. This is a key national strategic plan, which requires the integration of resources from multiple entities across the BJ-TJ-HE region, combining advantages, coordinating efforts, and ensuring proactive collaboration. The plan for integration also provides a modern vocational education serving society, meeting the needs of society, and deepening integration of industry and education, between schools and enterprises.

Using the "Industry-Education Collaboration Committee" as a platform, Tianjin City Vocational College has been active in the promotion of a coordinated vocational education across the three regions, as well as the development of elderly care resources, coordinated training of elderly care professionals, and collaborative research linking industry and education. It has exchanged useful experiences and successful practices, thereby playing a leading role in this field.

References

Chen, S. Q. (2000). Research review of China's elderly care model. *Population Studies, 3*, 30–36, 51.

Li, D. L. (2016). *Prospects for Hebei to undertake Beijing's elderly care industry under the coordinated development of the Beijing, Tianjin and Hebei region*. Hebei Normal University.

Mu, G. Z. (2002). Thoughts on China's policy on ageing. *Population Research, 26*(1), 43–48.

Wang, Z. Y., & Lv, X. J. (2015). Analysis of problems from the perspective of welfare pluralism in the provision of elderly care services in China. *China Labor, 04*, 72–78.

Wu, Y. Z. (2014). Report on the development of China's aged care industry. China Research Centre on Ageing, Beijing

Zhang, L. L. (2006). *Responding to an ageing population: construction and planning of a social pension system*. Social Sciences Academic Press.

Zhao, P. P. (2015). *Integration of the Beijing-Tianjin-Hebei region—Research on the coordinated development of the elderly care industry*. Hebei University.

Jian Fu is the Dean of the Social Undertaking Department and a major pioneer of elderly care and management in Tianjin City Vocational College, and office di-rector of the secretary of The Association of Industry and Education Cooperation for the training of aged Professionals in the Beijing, Tianjin, Hebei Region. He has got the master's degree of vocational education. The main research direction is the management of the aged enterprise and the training of the aged talents. He has hosted over the construction of the key major of service and management for the elderly in Tianjin during the 13th Five-Year Plan and Construction of Tianjin "High-quality Specialty Group connecting advantage Industry Cluster" Project. He has participated in the construction of Management of the nursing homes in the aged care and management which belongs to the teaching resource bank of the ministry of education. The achievements of professional construction have won the second prise of national teaching achievement and the special prise of teaching achievement of Tianjin.

Chapter 10
Age Simulation Suits—The First Step in Creating Empathy und Understanding for the Elderly

Andreas Lauenroth, Stephan Schulze, Birgit Teichmann, Kevin Laudner, Karl-Stefan Delank, and René Schwesig

10.1 Introduction

A recognisable increase in life expectancy may be caused by various advancements in medicine, education and technology. As such, people today live longer, which leads to ageing societies with a higher proportion of active years. The proportion of young people (under 20 yrs.) in Germany will decline from 18% (2015) to 16.2% by 2050. In contrast, the proportion of older people (over 60 yrs.) will grow from 15.4% (2015) to 17.5% (2050) and the proportion of those over 80 yrs. will increase from 5.8% (2015) to 13% (2050) (Statistisches Bundesamt, 2015). This development can also be applied to other Western industrial nations (Lavallière et al., 2016). In China, the number of people over the age of 65 has more than doubled since 1980. By contrast, the proportion of 65-year-olds and older in the total population doubled between 1950 and 2015 from 4.5 to 9.6% and for the year 2060 a tripling to 32.9% is expected. With a projected population of 1.3 billion in 2060, this corresponds to a share of approximately 430 million elderly (UN DESA, 2015).

Due to such developments, the number of people with physical limitations is increasing as well, such as other sensory or visual impairments, and reduced physical

---

A. Lauenroth (✉) · S. Schulze · R. Schwesig
Medical Faculty, Martin Luther University (MLU) Halle-Wittenberg, Halle, Germany
e-mail: lauenroth@univations.de

B. Teichmann
Network Ageing Research, Heidelberg University, Heidelberg, Germany

K. Laudner
College of Applied Science and Technology, Illinois State University, Normal, IL, USA

K.-S. Delank
Department of Orthopedic and Trauma Surgery, Martin-Luther University (LMU) Halle-Wittenberg, Halle, Germany

© The Author(s) 2025
Deutsche Gesellschaft für Internationale Zusammenarbeit (GIZ) GmbH et al., (eds.), *Sustainable Aging*, https://doi.org/10.1007/978-3-662-69139-7_10

151

strength (Lavallière et al., 2016). The physical impairments also mean that everyday activities are becoming increasingly complicated and difficult for older individuals. Especially for the social, economic and medical fields, the ageing process and its consequences seem to be a huge challenge. Although this demographic change brings up many challenges, the elderly (60+) are also an interesting target group for trade, industry and service. The insight into the experiences and an increasing understanding of behaviours and needs of older people can be very helpful and effective for the younger generations (Fisher & Walker, 2013).

The increase in our ageing society will entail a subsequent increased demand for employees in the nursing profession. In addition, there will also be more occupational activity fields that work with the elderly. Therefore, it is important to create a basis for increased empathy for older people, especially in education and care (Lauenroth et al., 2017). One way to build understanding and empathy for the elderly, at work and in everyday life, is the use of age simulation suits (see Fig. 10.1). Age simulation suits are primarily intended to draw attention to the age and ageing process. As early as the 1970s, age simulation suits were used to alert and educate doctors and medical staff regarding the experiences of older individuals. In 1995, the first commercially used age simulation suit was introduced and distributed. Today, there are several providers that offer and market age simulation suits.

Educational models describe experimental and practice-oriented design for sustainable learning as meaningful (Braude et al., 2015; Brunero et al., 2010). When using a variety of simulation methods, the most realistic situation should be presented

Fig. 10.1 The age simulation suit in daily life situation. *Source* Lauenroth

to convey empathy and understanding of ageing. So far, the use of age simulation suits is sparse and serves often the media staging too.

There are only a few scientific studies and experiments in which the use of these suits has been experimentally investigated and discussed (Kullman, 2016; Lauenroth et al., 2017; Lavallière et al., 2016).

10.2   Methodology

With an age simulation suit, such as the **GER**ontological **T**est suit (GERT), it may be possible for younger individuals to understand age-related limitations of the elderly, such as lack of strength, joint stiffness, and sensory limitations. The age simulation suit consists of a weighted vest (9,0 kg), wrist bandages for elbows and knees as well as weighted cuffs for the wrists and ankles (1,5 kg each) (Fig. 10.1). The sensory limitations as we age (e.g. vision and hearing impairments) are simulated by special glasses and hearing protection (Fig. 10.2). In addition, different components are used to represent spinal compression, muscle atrophy, reduced balance and stability, as well as decreased joint mobility. The suit was designed as a tool to represent the natural causes of physical limitations based on human anatomy. Interaction of all these physical components of the suit helps to create a realistic experience of the ageing process for the suit user.

Our study attempted to determine if these suits provide the same physical limitations experienced by older adults. More specifically, we were interested in determining how many "additional years" the suits add to the individuals experience. Volunteers walked at a self-selected pace both with and without an age simulation suit. Gait parameters were recorded with a gait analysis system and compared with each other. The comparison of gait parameters both with and without suit provided valuable conclusions to the gait patterns in older populations. We also compared the gait parameters of younger people while wearing the suit with the gait pattern of older people who were not wearing the suit. Differences between age groups were tested using a univariate analysis of variance with age as the independent factor (Bortz & Schuster, 1999).

10.2.1   Results

For a realistic interpretation of the perceived age while using these suits our study attempted to objectify these physical limitations. To this end, 178 healthy volunteers (18–85 years, avg. 50.4 ± 16.4 yrs.) were enrolled. All subjects completed various test conditions with and without the age simulation suit. Among other things, spatio-temporal gait parameters such as gait velocity, step length, and base width were examined to evaluate the influence of the age simulation suit on gait performance.

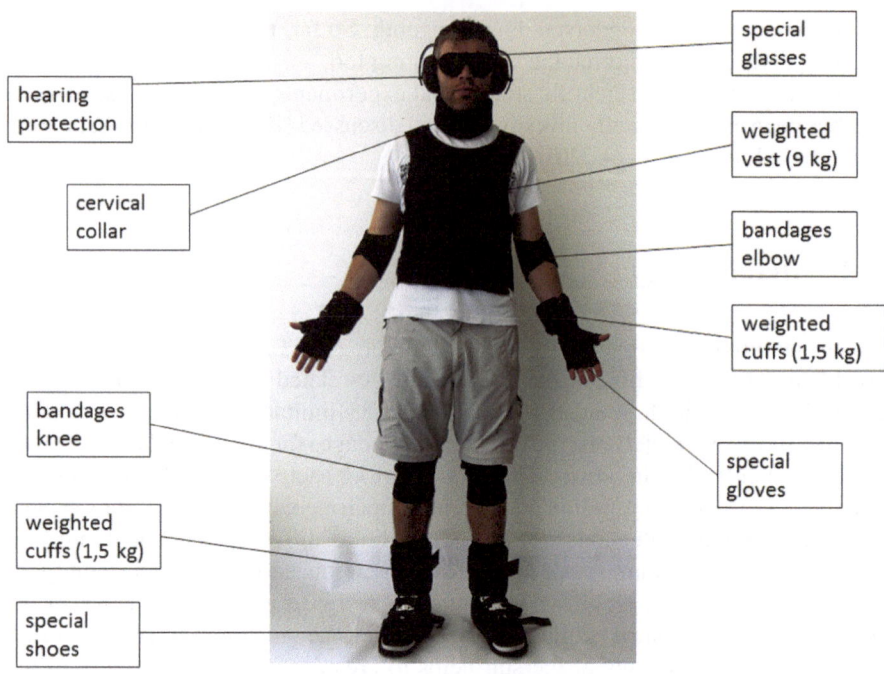

Fig. 10.2 Components of the age simulation suit—hearing protection, spectacles, cervical collar, weight vest, bandages, weight cuffs and gloves. *Source* Lauenroth

Subsequently, the gait performance of younger subjects while wearing the suit was compared with the gait performance of the elderly subjects without a suit.

After evaluating the data, the authors conclude that the simulation suit provided a perceived physical age of an additional 20–25 years (Fig. 10.3). Especially during gait velocity and step length, which showed that velocity while wearing the suit in the 40–49-years-old age group corresponded to the velocity of the 60–69-years-old age group without a suit. However, this additional age varies and is dependent of factors such as lifestyle, constitution, physical activity, etc.

The added age should therefore be interpreted as a tendency and is not entitled to be a reliable simulation. With an "additional" age of 20–25 years, in the field of medical and nursing education, the suit is not only a pedagogical tool for students and trainees, but also suitable for giving middle-aged professionals (35–60 years) an idea of how older people feel. This could be done through further education or workshops.

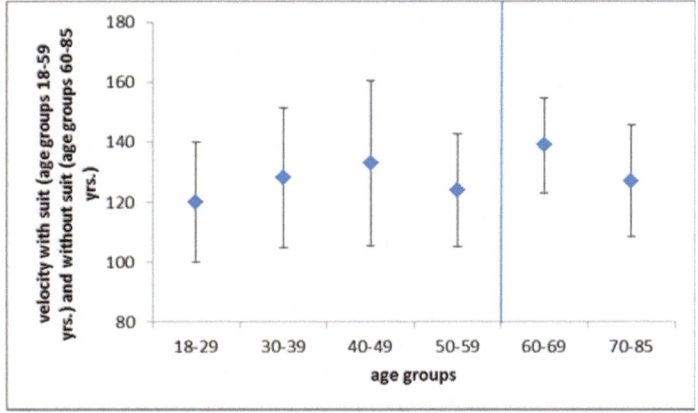

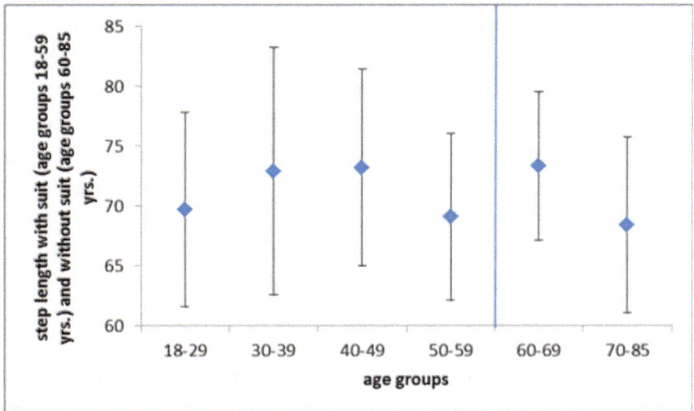

Fig. 10.3 Comparison of gait velocity (above) and step length (below) in four younger groups (18–59 yrs.) with age simulation suit and both older groups (60–85 yrs.) without age simulation suit—means and standard deviation (modified according to Lauenroth et al., 2017, Fig. 10.2)

10.2.2 Discussion/Conclusions

The user can generally benefit from experience with the suit wherever he works or communicates with older people. This applies to all areas of medicine and care, as well as areas where older people are users or customers.

Especially for education and training in the field of nursing, the age simulation suit can be an effective and convincing tool to give young people insights into the feelings and needs of older people. Similarly, activities of daily living (ADLs) with respect to the elderly can be experienced with the suit, which leads to the empathy education among younger people. For example, it is easier to understand why older people move slower in supermarkets or spend more time at the cash register looking for small change (Fig. 10.2).

In fact, there are already approaches in which public transport drivers were trained to consider the accelerating and braking reduced responsiveness of older people. These precautions may help avoid falls and serious consequences for those affected and recourse claims against the company. For the development of assistance systems in vehicles, use of the suit is equally useful for architects of senior-friendly apartments and outbuildings (such as parks). Likewise, providers of senior recreational courses (dancing, sports, and gymnastics) can learn why older people move slower and more cumbersome. This impulse is profitable in almost all areas of social life and for both older and younger people.

Another example can be found in the manufacturing industry. Many work processes are planned and carried out independently of age. It can be crucial if a younger or older employee is part of such a process. A younger employee may work more physically and benefit from his/her physical strength. In return, an older worker tends to work much more economically, as he/she can draw on a wealth of experience. In this context, there can be misunderstandings and tensions within the workforce. This could be addressed by making younger people aware of the physical expectations and needs of older colleagues.

The older worker can understand the benefits of youth since he/she has experienced it him/herself. Conversely, a young person has no experience with what it is like to be older. At this point, an age simulation suit would be helpful to allow some insight into these experiences. Proceeding with such impressions of the younger workers, it is advisable to work in Tandems young/old, to combine the benefits of both generations, so as not to lose the experience of age-related leaving. The result would be consistently high productivity, an understanding of each other and a direct transfer of existing knowledge.

Compassionate treatment of older people is particularly important when older people need the help and care of younger people. This situation is particularly common in medical treatment and follow-up care. Many patients and people in need of care are both physically and mentally alert to medical personnel or care-givers identifying their needs and acting appropriately. Also in this case, tandems or working groups young/old as well as the use of age simulation suits are recommended, as in this case not only the medical staff learns from each other but will also indirectly benefit the people who are in need of help. This is a very favourable win–win situation here.

There is some evidence in the literature that simulation programmes can be useful to both health care professionals and nursing professionals in developing empathy and understanding of the elderly (Tremayne et al., 2011). The development of age-related products and services is becoming increasingly important due to demographic changes, ultimately serving the elderly as well as society. For engineers and developers, this creates a field of activity in which existing products can be further developed or new products and services can be created. Overall, numerous users could benefit from these experiences. One of the first scientific studies dealing with the application orientation of age simulation suits, showed that this simulation can at the very least give a first impression of the everyday activities of older people (Lavallière et al., 2016).

Our research has shown that these suits can accurately simulate the gait pattern of individuals 25–30 years older (Lauenroth, 2013; Lauenroth et al., 2017).

Thus, for the first time, the quantifications of these age simulation suits can be objectified. The perceived age experienced by the suit is influenced by lifestyle parameters such as health status, physical activity, diet, interpersonal relationships, as well as gender and anthropometric parameters. The practical experiences gained through self-experience are thus linked to real behaviour and not merely a theoretical mediation of knowledge at a superficial level (Brunero et al., 2010; Halpern, 2003; Ross et al., 2012).

Due to the demographic change, the challenges and possibilities of the increasing ageing of society should be addressed at an early stage. These suits may make it possible to adequately respond to the unforeseeable consequences of ageing. The growing number of elderly people will mean that it will be necessary to adapt to changing needs. Particularly in large countries with large populations, the effects of ageing could be disproportionately experienced. In countries such as China or India, it is particularly important to face the challenges at an early stage. In addition to health care, labour market policy, housing and living environment design, public infrastructure, that also includes an educational structure, helps young(er) people, among others, to be prepared for life in an ageing society. Part of this education should also be a geriatric care area, which deals specifically with the service and treatment of people in need of care.

A well-designed and structured care for the elderly would respond to the development of ageing populations. Young nursing trainees thus can do their part to make life easier for older generations and to deal adequately with increasing infirmity and support (medical and social). This would benefit both society and the elderly directly.

The use of realistic illustrative materials could be used early on to draw attention to the area of geriatric care in schools and to present a job profile that will be needed in the future.

10.3 Summary

An age simulation suit allows younger people self-awareness to experience the physical limitations of older age and the ageing process. The experienced simulation components result in the positive influence of individuals perspectives regarding the behaviour and needs of older people. A study using investigating the suit´s effect on human gait showed that the age-related changes can be adequately depicted with the suit. This gives us the opportunity to experience how certain restrictions have a physical effect with an age difference of 25–30 years. This may be of particular interest for designers, executives and engineers of age-related products and services to meet the needs of demographic change.

The benefit of the presented components is complex. On the one hand, these findings can be used for one's own professional activity and dealing with one's own ageing. On the other hand, older people benefit indirectly by increasing their empathy towards the own perspective as well as by providing an improved supply of ageing-friendlier products and services.

However, the age simulation suit and its use should not fulfil excessive expectations. While "instant ageing" temporarily experiences physical limitations of age, it cannot replace a personal experience of age. In addition to that, various "coping strategies" and personal lifetime experiences, such as pain, cannot be simulated. As such, the age simulation suit represents only selected physical limitations. In addition, both the health of the participants (e.g., cognitive limitations) and physiological and genetic factors influence the subjective experience of the simulation. The experienced age difference of 25–30 years represents an average recommendation value of a larger sample and may differ for individual users. The effect of an age simulation suit is not exhaustive. Rather, they are intended to reflect general physical limitations and provide first/new insights into the challenges and effects of ageing.

Competing Interests All authors declare that they have no competing interests.

References

Bortz, J., & Schuster, C. (1999). *Statistics for human and social sciences* (7th ed.) [book in German]. Springer

Braude, P., Reedy, G., Dasgupta, D., Dimmock, V., Jaye, P., & Birns, J. (2015). Evaluation of a simulation training programme for geriatric medicine. *Age and Ageing, 44*(4), 677–682.

Brunero, S., Lamont, S., & Coates, M. (2010). A review of empathy education in nursing. *Nursing Inquiry, 17*(1), 65–74.

Fisher, J. M., & Walker, R. W. (2013). A new age approach to an age old problem: Using simulation to teach geriatric medicine to medical students. *Age and Ageing, 43*(3), 424–428.

Halpern, J. (2003). What is clinical empathy? *Journal of General Internal Medicine, 18*(8), 670–674.

Kullman, K. (2016). Prototyping bodies: A post-phenomenology of wearable simulations. *Design Studies, 47*, 73–90.

Lauenroth, A. (2013). Influence of an ageing suit on gait performance in persons aged from 20–71 years. *The Gerontologist, 53*, 325–326.

Lauenroth, A., Schulze, S., Ioannidis, A., Simm, A., & Schwesig, R. (2017). Effect of an age simulation suit on younger adults' gait performance compared to older adults' normal gait. *Research in Gerontological Nursing, 10*(5), 227–233.

Lavallière, M., D'Ambrosio, L., Gennis, A., Burstein, A., Godfrey, K. M., Waerstad, H., Puleo, R. M., Lauenroth, A., & Coughlin, J. F. (2016). Walking a mile in another's shoes: The impact of wearing an age suit. *Gerontology and Geriatrics Education*, 1–21.

Ross, A. J., Anderson, J. E., Kodate, N., Thomas, L., Thompson, K., Thomas, B., Key, S., Jensen, H., Schiff, R., & Jaye, P. (2012). Simulation training for improving the quality of care for older people: An independent evaluation of an innovative programme for inter-professional education. *BMJ Quality and Safety, 22*(6), 495–505.

Statistisches Bundesamt. (2015). Bevölkerung Deutschlands bis 2060—Ergebnisse der 13. Koordinierten Bevölkerungsvorausberechnung. Retrieved from: https://www.destatis.de/DE/Publik ationen/Thematisch/Bevoelkerung/VorausberechnungBevoelkerung/BevoelkerungDeutschla nd2060Presse.html [03.04.2017].

Tremayne, P., Burdett, J., & Utecht, C. (2011). Simulation suit aids tailored care. *Nursing Older People, 23*(7), 19–22.

UN DESA, United Nations, Department of Economic and Social Affairs. (2015). *World population prospects: The 2015 revision.* Retrieved from: The World Population Prospects: 2015 Revision | Latest Major Publications, United Nations Department of Economic and Social Affairs [20.05.2020].

Andreas Lauenroth studied Sports Science and Politics at the University of Halle-Wittenberg (MLU) from 1998 to 2003. He finished his doctorate in 2008 on the treatment of balance disorders. From 2009 to 2015 he worked at the University of Heidelberg in the field of Sports Science/Aging Research. As a visiting researcher, he also conducted research in Tübingen at the Hertie Institute for Clinical Brain Research (2010–2012) and at the Massachusetts Institute of Technology (MIT) in Boston (2011). His research interests include the diagnosis and treatment of balance disorders, fall prevention, the development of methods for measuring posture, and aspects of aging research with regard to gait and posture (e.g. use of age simulation suits). He also actively supports the transfer of knowledge from research to therapeutic practice. He is currently working as a research assistant in the field of aging, diagnostics and sports medicine at the Medical Faculty of MLU Halle-Wittenberg. In addition, as a senior project manager, he is involved in future innovations in the healthcare sector.

Stephan Schulze studied Sport Sciences from 2001 to 2007 at the University of Halle-Wittenberg. Afterwards he worked as Research Associate in field of performance testing and exercise sciences at the institute of performance diagnostics and health promotion and as a scientific employee and teacher in the field of Exercise Science and Sports Medicine at the sport institute of the Martin-Luther-University Halle-Wittenberg. He finished his doctorate in 2015 on training and performance testing in normobaric hypoxia. His research interests include performance physiology in individual and team sports as well as biomechanical aspects of posture, gait and running patterns. He is also engaged in proceeding knowledge transfer from science to partners from competitive sports. He is working as a Research Associate in field of Biomechanics and Sports Medicine at the medical faculty of the MLU Halle-Wittenberg.

Birgit Teichmann studied biology at the University of Bonn and the University of Heidelberg from 1991 to 1996. In 2000, she completed her doctoral thesis at the German Cancer Research Center in Heidelberg. She has been working at the Network Aging Research as scientific manager since 2007, the year the organization was founded. She also holds Master's degrees in "Managing Aging and Chronic Diseases" from the Hellenic Open University (M. Sc. 2020) and in Bioethics from the Open University of Cyprus (M.A. 2023). Her research interests include programs for aging societies, dementia-friendly hospitals, the effect of bilingualism on the onset of dementia, and dementia awareness programs for informal caregivers.

Kevin Laudner is a certified athletic trainer with over 20 years of clinical experience working with a variety of patient populations. A Professor and Associate Dean of Research in the College of Applied Science and Technology at Illinois State University in the United States, his research focuses on the prevention and care of shoulder pathologies and manual therapy techniques in the evaluation and treatment of various orthopaedic injuries. He is currently collaborating with several professional sport teams to efficiently identify and treat upper and lower extremity pathologies in professional athletes. He holds a Doctor of Philosophy degree in Rehabilitation Science (2004) with an emphasis in Sports Medicine, a Master of Science degree in Health and Human Performance (1999), and a Bachelor of Arts in Kinesiology (1997) with an emphasis in Athletic Training. He has more than 50 peer-reviewed journal publications and routinely lectures at international and national professional meetings.

Karl-Stefan Delank studied Medicine at the universities of Gießen and Munich from 1988 to2004. In 2007, he decided to specialise in the field of "functional changes within backpain". He has worked as a medical doctor in the university hospitals of Bochum, Mainz and Cologne from 1995 to 2011. Since then he has been working as the Medical Director of the Department of Orthopaedic and Trauma Surgery, Martin-Luther University Halle-Wittenberg in Germany.

René Schwesig currently serves as the Head of Laboratory of Sport Medicine and Experimental Orthopaedics at the medical faculty of the Martin-Luther University Halle-Wittenberg. He studied Sport Science from 1993 to1997 at the University of Halle-Wittenberg. He worked afterwards as a tutor in the field of Bio-mechanics at the local institute of Sports Science. He completed his doctorate in 2001, for which he focused on the subject "Therapy of back pain". Since 2006, he began to specialise in the topic of "postural regulation in life span". In 2016, he was nominated as an outstanding professor at the medical faculty of the University Halle.

Chapter 11
Project Case: Promotion of Dual Vocational Education in Panjin—*Geriatric Nursing at Panjin's Vocational and Technical College*

Li Lin and Annika Fründt

Facts

Project period: 03/2017–08/2020

Target group: Teachers and students in Panjin Vocational and Technical College

Commissioner: Panjin Vocational and Technical College

Project activity location/country: Panjin, Liaoning Province, China.

11.1 Social Challenge Behind the Project

The availability of highly qualified specialists, with technical as well as practical skills, is crucial for the long-term economic success of the north-eastern Chinese city of Panjin, as ranked both nationally and internationally. A functioning (geriatric) nursing care system is necessary in China to respond to demographic developments. Thus, Panjin is one of the first cities in Northeast China to pilot a dual vocational training system close to the German model. The combination of technical knowledge and practical skills makes graduates from the German vocational system highly skilled sought-after specialists. As a pilot institution, Panjin Vocational and Technical College (PVTC) is introducing a modern vocational training campus in line with international standards, delivering work-oriented training including in geriatric nursing, which will see highly qualified skilled worker's graduate.

L. Lin (✉) · A. Fründt
Deutsche Gesellschaft Für Internationale Zusammenarbeit (GIZ) GmbH, Bonn, Germany
e-mail: lin.li@giz.de

© The Author(s) 2025
Deutsche Gesellschaft für Internationale Zusammenarbeit (GIZ) GmbH et al., (eds.),
Sustainable Aging, https://doi.org/10.1007/978-3-662-69139-7_11

11.2   Project Approach

The structure of the newly introduced dual vocational programme is being systematically implemented in 6 areas of action. These include joint development and revision of existing curricula between German experts and the PTVC, and the corresponding teaching and learning materials. Furthermore, with the new curricular set up, the learning didactics and exam preparation is being revised, and quality management introduced.

A crucial element of German vocational training is the practical training in the workplace. Hence, a cooperation framework has been put in place between the respective hospitals and the college, to ensure the developed programme is tailored to the needs of the profession and that the trainees are fully operational after graduation.

The pilot class consists of 24 students, with most girls aged 18–19, although some boys have started opting for nursing education. All the students have completed internships in hospitals and social care homes in the city of Panjin.

11.3   Value Added

The project aims to introduce a cooperative and competence-based training model, to deliver sustainable training, as well as analytical and problem-solving skills. The (geriatric) nursing professionals gain expertise in both clinical and geriatric care and can use their nursing skills to carry out the entire care process for sick as well as elderly people, a growing demographic. The training responds to needs in various clinical and nursing institutions in the fields of nursing, prevention, rehabilitation and care. This provides training with a broad occupational field as well as rendering a necessary service to society.

11.4   Lessons Learnt

For graduates of the dual system to match the needs of employers appropriately, it requires companies and hospitals to continue to coordinate and express their needs to the college. Only with their active participation, can a tailored programme for systematic training be set up. However, the engagement of companies and hospitals is not yet at the desired level. Hence, these partners need to be included in the project further upstream, with their engagement fostered in a multi-level manner to be able to deliver adapted and sustainable training, which would benefit their long-term interests.

11.5 Ideas for the Future

In addition to the hospitals and cares homes, outpatient nursing services and assisted living are of crucial importance. There are more and more private nursing homes for the elderly in China, where the qualified nursing staff can organise leisure activities for the elderly in addition to hygienic care. This offers new challenges and opportunities for the dual vocational training in geriatric nursing.

Li Lin holds a master's degree in translation and interpreting of Beijing International Studies University. Currently, she works for the Deutsche Gesellschaft für internationale Zusammenarbeit (GIZ) GmbH as project officer and interpreter. She is responsible for planning and implementation of TVET projects in Panjin Vocational and Technical College and Jincheng Institute of Technology. She has professional experience in interpreting at a TCM clinic in Germany and at government office in China.

Annika Fründt is a project manager in the Asia team at GIZ International Services. She holds a bachelor's degree in European Studies from Maastricht University, and a master's degree in International Relations from Leiden University, Netherlands. While her research targeted societal developments in the international context, her focus shifted when she began her career in project management in international coop-eration. After working for the international cooperation consultancy AHT GROUP AG between her B.A. and M.A., she joined GIZ International Services in 2017. She oversees a variety of different TVET projects in China, which all aim to de-velop a systematic, tailor-made vocational training approach, inspired by the German vocational training system and fitting the needs of the Chinese graduates and employers. The projects are carried out in cooperation with a variety of lev-els, including individual technical vocational colleges, city governments, as well as a Provincial Education Department.

Chapter 12
Project Case: Establishing Elderly Care System in China—Sustainable Improvement of Situation for the Elderly

Mingming Wang and Sabine Porsche

Facts

Project period: 12/01/2016–11/30/2019

Target group: Government Officials, Elderly Care Professionals, Colleges, Senior Citizens

Commissioner: Federal Ministry for Economic Cooperation and Development (BMZ)

Partners: RENAFAN GmbH

Countries of project activities: China.

12.1 Social Challenge Behind the Project

As the most populous country in the world, China has been experiencing demographic change for nearly two decades due to country's the plummeting birth rate and a rise in life expectancy. According to World Bank data, China's total fertility rate was 1.62 in 2016, far behind the global average of 2.43. The lingering consequences of the previous strict family planning policies combined with increasing longevity have meant that China will have to confront the challenges of a rapidly ageing population at a relatively early stage of its economic development, which limits the financial resources for supporting the elderly. The progressive ageing of China's population could also yield major consequences upon the resilience of the national economy. Indeed, it is questionable to what extent China has prepared itself for the economic

M. Wang (✉) · S. Porsche
Deutsche Gesellschaft Für Internationale Zusammenarbeit (GIZ) GmbH, Bonn, Germany
e-mail: mingming.wang@giz.de

© The Author(s) 2025
Deutsche Gesellschaft für Internationale Zusammenarbeit (GIZ) GmbH et al., (eds.),
Sustainable Aging, https://doi.org/10.1007/978-3-662-69139-7_12

implications of this demographic phenomenon. Several policy approaches could help China better prepare itself for the financial burden imposed by the demographic change. For example, the introduction of a comprehensive elderly care system has the potential to sustainably strengthen both the economy and society. The Chinese government has included the introduction of an elderly care system as one of the main tasks of the 13th Five-Year Plan for Economic and Social Development.

12.2 Project Approach

By fostering cooperation with public and private partners, the project sought to support the Chinese government to develop innovative concepts for the country's elderly care system. A core goal of the project was to develop a geriatric care concept which emphasises the implementation of quality standards for geriatric care. The project followed a multi-level approach specifically designed to address the complex topic of elderly care. On the project level, training curricula for elderly care are developed based on the German Dual Vocational Training Model with local vocational schools. On the societal level, the project's awareness-raising campaigns strived to improve public image of the elderly care profession. On the institutional level, the project provided consultancy regarding quality standards for elderly care services. It also provided innovative models of community elderly care, such as assisted living and ambulatory care. In addition, the project cooperated closely with political institutions to support the policy process as well as the development of elderly care models and relevant legislature in the future. The project partner RENAFAN GmbH is a German company which contributes to the project by conducting trainings and providing consultancy support on elderly care services.

12.3 Value Added

Through intervention in education and training, technical and public dialogues to improve elderly care quality to reach international standards, dissemination of knowledge and the training of multipliers, the project significantly contributes to the development of solutions for the multidimensional problems posed by this demographic shift. The project was also involved in reforming government agendas, plans and implementing pilot projects to improve the lives of the elderly in China. Furthermore, the project supports the German Chinese cooperation in the health sector on "Healthy Ageing", which was agreed upon during the 5th Sino-German Government Consultation in July 2018.

12.4 Lessons Learnt

The project shows that it is important to involve all relevant stakeholders (political and non-political) to successfully establish the basic elderly care system in China. By involving different political agencies in joint project activities, this approach ensures the political exchange and greatly contributes to the project's success. In addition, continuing exchange with other international partners and the joint activities would prove extremely beneficial to boosting understanding of the challenges related to the ageing population and, therefore, the development of relevant solutions.

Besides that, capacity building is of great importance to raise awareness. Within three years more than 5000 leading professionals of local governments, medical colleges and elderly care facilities attended trainings in 13 provinces. Topics were quality management in elderly care facilities, integrated care and medical services as well as dementia. The German dual vocational education system to train elderly care nurses was adapted to local standards and implemented in three medical colleges in China. Around 100 students started their three-year vocational education training within the project.

12.5 Ideas for the Future

As new discussions emerge on the topic of China's elderly care services, such as long-term care insurance, it's clear that several of these areas could benefit from German expertise. Therefore, this project will also strive to support the further development and testing of the training curricula in correspondence with the available alternative elderly care models. Additionally, within the project framework, new concepts will be developed, while formats and tools, designed to promote the social acceptance of working as a caretaker for elderly people, will be further supported.

Mingming Wang works as Senior Project Manager in the Cluster of Labour and Social Security at GIZ China. With several years working experiences on public health and social protection, her main responsibilities are in the field of cooperation with private sector on elderly care, public health and social protection topics. After the implementation of the public-private partnership project "Introduction and Dissemination of basic elderly care system in China", she is now focusing on the possibility of cooperation between China, Germany and Africa in the field of health care management.

Sabine Porsche graduated in Cultural Anthropology from Philipps-University in Marburg. She joined Deutsche Gesellschaft für Internationale Zusammenarbeit (GIZ) in Beijing in 2017, where she heads the team Society and Labour. Topics of interest are elderly care and education. From 2007 to 2016 Sabine held different positions at Tongji University in Shanghai. She held the position as German Vice-Director of the Sino-German University of Applied Sciences and besides established the DAAD Career Academy.

Chapter 13
Project Case: Triple Win Project—Fair and Sustainable Recruitment of Nurses

Maja Bernhardt and Sonja Alves Luciano

Facts

Project period: Since 2013 ongoing.

Target group: Qualified personnel in the field of nursing.

Commissioner: Private sector.

Partners: International Placement Service (ZAV) of the German Federal Employment Agency (BA) and the labour administrations of countries of origin.

Countries or places of project activities: Germany, Bosnia-Herzegovina, the Philippines, Indonesia, India/Kerala and Telangana and Tunisia.

13.1 Social Challenge Behind the Project

Germany's nursing sector is already feeling the impact of a significant shortage of nurses today. At present, vacancies outnumber the amount of qualified job seekers on the job market. According to expert estimations, the nursing sector will need 500,000 new nurses by 2030 (Bertelsmann Stiftung, 2012: 10f). Demographic changes in the country will exacerbate this situation in the medium and long term. By contrast, in Bosnia and Herzegovina, Indonesia, India/Kerala and Telangana, the Philippines and Tunisia, there is a surplus of qualified experts that cannot be absorbed by the local labour markets. This has resulted in a high level of unemployment among nurses in these countries.

M. Bernhardt · S. A. Luciano (✉)
Deutsche Gesellschaft Für Internationale Zusammenarbeit (GIZ) GmbH, Bonn, Germany
e-mail: sonja.alves-luciano@giz.de

© The Author(s) 2025
Deutsche Gesellschaft für Internationale Zusammenarbeit (GIZ) GmbH et al., (eds.),
Sustainable Aging, https://doi.org/10.1007/978-3-662-69139-7_13

13.2   Project Approach

The BA's International Placement Services (ZAV) and GIZ have established a joint project for the placement of qualified nurses with German companies (clinics and nursing homes). It is financed by the German companies that pay a service fee of EUR 7,900 EUR gross to GIZ for the preparatory measures and accompaniment in Germany. The migration is based on placement agreements between the ZAV and the labour administrations in the partner countries. Moreover, the project intends to prevent brain drain in taking into consideration those countries, which have a surplus of well-trained nurses. Triple Win manages the selection of qualified applicants in its partner countries, as well as their linguistic and professional preparation. Nurses and clinics are accompanied during the integration and the process of recognition of the exam in Germany.

The migration from the participating countries of healthcare personnel who can demonstrate a suitably high standard of training presents a wide range of opportunities for everyone involved and generates threefold benefits ('triple win'):

1. Pressure is eased on labor markets in the countries of origin. Migrants' remittances provide a developmental stimulus in their countries of origin.
2. Migrating in this way provides the nurses with the chance to improve their future prospects
3. The shortage of nurses in Germany is alleviated.

13.3   Value Added

From the beginning of the project until now (June 2024), more than 6,200 nurses have been placed with German employers, in clinics, geriatric care homes and outpatient services. Over 2500 of them already work in Germany while the remaining nurses are currently in several stages of departure (linguistic and professional preparation). Altogether, Triple Win worked with about 400 employers since its beginning. After eleven years, the programme is now well established and has gained recognition also at international level. It has been commended as best practice by the International Organisation for Migration (IOM) and the Public Services International (PSI).

13.4   Lessons Learnt

Migration can be considered successful if both professional and social integration are achieved. Therefore, preparation and accompaniment both for nurses and the employers in Germany is crucial. Bureaucratic obstacles have to be overcome. The nurses themselves recognise the benefits of the project – despite the initial challenges of learning German or getting their qualifications and training recognised. They feel well integrated and have grown both professionally and personally. Many of them have already completed advanced training, laying the foundations for a successful future in Germany.

13.5   Ideas for the Future

New partner countries as well as new possible occupational categories could be made accessible. Furthermore, to maintain on the road of managing migration in a fair and sustainable way, Global Skill Partnerships could be the next step as a form of partnership. The Global Skill Partnership entails an advanced cooperation between migration countries and origin countries agreeing on investing in education and training in both countries. Therefore, GIZ was commissioned in the beginning of 2020 by the German Federal Ministry of Health (BMG) to build up cooperation between a German university hospital and a university in Mexico, the Philippines and Brazil to build up qualification measures for nursing students that have the potential to shorten the recognition process in the migration country but also profit from collective professional learning. Also, the Bertelsmann Foundation (BST) commissioned a same approach for one more cooperation with a university in the Philippines.

Reference

Bertelsmann Stiftung (Ed.) (2012). *Themenreport „Pflege 2030".* Retrieved from https://www.ber telsmann-stiftung.de/fileadmin/files/BSt/Publikationen/GrauePublikationen/GP_Themenrep ort_Pflege_2030.pdf. Accessed May 21, 2020.

Maja Bernhardt served as the Project Leader for the "Triple Win Project" on behalf of GIZ International Services. She has previously worked in different areas of healthcare and social work such as by promoting health projects for various target groups. She studied Social Sciences and Social Work at the University of Kassel in Germany

Sonja Alves Luciano works as a Project Lead for the Triple Win Project on behalf of GIZ International Services. She previously worked as Placement Officer for the International Placement Agency (ZAV) of the Federal Employment Administration (BA) of Germany, which is also within the Triple Win Project. She studied Social Sciences and Intercultural Relations (BA) at University for Applied Science in Fulda (Germany) and Human Geography (MA) at Goethe University in Frankfurt (Germany). During her studies, she undertook research in the recognition processes of foreign skills in Germany as well as the personal and professional possibilities of migrants in Germany.

Chapter 14
Project Case: Secure Skilled Labour in Elderly Care—A Sustainable German-Vietnamese Approach

Florian Krins

Facts

Project period: First project, Elderly Care (2012–2016), second project, Nursing (2016–2019)

Target group: Graduates of Vietnamese nursing colleges with an interest in a dual vocational training in nursing or elderly care in Germany; German employers

Commissioner: Federal Ministry of Economics and Energy (BMWi) (Today: Federal Ministry of Economic Affairs and Climate Action, BMWK)

Partners: Ministry of Labour, Invalids and Social Affairs (MoLISA)—Department of Overseas Labor (Vietnam)

Countries or places of project activities: Vietnam/Germany.

14.1 Social Challenge Behind the Project

Germany is facing an alarming shortage of qualified nurses in the health and geriatric care sector. Experts estimate that, due to demographic change, the number of people requiring care in Germany will increase from the current level of about 2.3 million to approximately 3.4 million in 2030. The Federal Employment Agency (BA) have already observed a severe shortage of fully qualified (geriatric) nurses and are warning of an impending acute shortage of nursing staff. This shortage cannot be covered in the medium or long term by nurses in Germany itself. At the same time, many young people from abroad are keen to take up training and a subsequent period of work spent

F. Krins (✉)
Deutsche Gesellschaft Für Internationale Zusammenarbeit (GIZ) GmbH, Bonn, Germany
e-mail: florian.krins@giz.de

© The Author(s) 2025

173

Deutsche Gesellschaft für Internationale Zusammenarbeit (GIZ) GmbH et al., (eds.),
Sustainable Aging, https://doi.org/10.1007/978-3-662-69139-7_14

in Germany. Vietnam has a very young population yet falls far short of employing its entire potential workforce in its labor market.

14.2 Project Approach

From late 2013 up until end of 2019, a total of 350 young people from Vietnam have either completed or are currently undertaking training work as nurses or care assistants for the elderly. After completing a 13-month language course at the Goethe-Institute in Hanoi, the participants receive training in small groups at care homes and hospitals in German pilot regions. In addition to learning German, the language course also provides a class on specialist terminology and an intercultural programme to prepare the participants for working in a country so far away from home and immersing themselves in a different culture.

The goal of the BMWK-funded project was creating greater awareness among German employers to render the recruitment process more when it comes to hiring foreign skilled workers in the nursing and care sector. Both parties should benefit from this scheme: the programme candidates gain an opportunity to work and train in Germany for the long-term; German hospitals and elderly homes can acquire the qualified trainees whom they urgently needed.

14.3 Value Added

The project is pursuing the "triple win"-strategy which generates threefold benefit for the migrants as well as the countries of origin and of destination. (1) Creating fair conditions in the participants' migration process and delivering opportunities for their personal and professional development is a central focus of the project. (2) Furthermore, the project tries to relieve pressure on the Vietnamese labour market. The participants' remittances also possibly contribute towards the enhancing of the local structures. (3) Finally, the German health and care sector benefits from the sustainable acquisition of qualified personnel and may even use this experience as a model for recruiting foreign qualified professionals in the future.

14.4 Lessons Learnt

Following the programme implementation, there were two key lessons learnt. Firstly, regarding the participants, a good level of German language skills is important not only for aiding their integration process in Germany but also for their successful completion of the vocational training course. In addition, it is very important to offer close support to the German employers during the integration process of trainees.

This support not only includes assistance in official processes (e.g. application for residence permit) but also in the sensitisation of the German colleagues by offering intercultural trainings.

14.5 Ideas for the Future

Because of the positive feedback received by the BMWK-funded model project, GIZ has completed the implementation process of a follow-up project with Vietnam which builds on the gained experiences. As the new approach is fully financed by German employers, it is implemented by GIZ International Services, the taxable business area of GIZ, which can be commissioned directly by the private sector. It´s running under the roof of the "Triple Win"- programme, the joined initiative of GIZ and International Placement Service of German Federal Employment Agency. Therefore, the BMWK-funded project reached its goal of sensitising the German health sector to enable a fair and sustainable recruitment process of personnel from third countries and transfer a model project financed by the public to the private sector.

Florian Krins graduated from Hamburg University with a Master of Arts degree in Sinology, Political Science and Modern German Literature in 2008. He first worked as a freelancer focusing on translation and public relations. In 2010, Florian joined Deutsche Gesellschaft für Internationale Zusammenarbeit (GIZ) and has been working for the organisation ever since. In 2013, he began specialising in the topic of legal migration and integration of qualified employees. Since 2016, he has worked as the Senior Project Manager for project that entails recruting personnel from Vietnam to train as nurses in Germany.

Chapter 15
Project Case: Sino-German Platform for Precision Oncology—Tailor-Made Trainings for Medical Professionals

Sabine Porsche and Shuo Cui

Facts

Project period: October 2020–September 2023

Target group: Chinese medical professionals applying Precision Oncology

Commissioner: Federal Ministry for Economic Cooperation and Development (BMZ)

Implementer: The Deutsche Gesellschaft für Internationale Zusammenarbeit (GIZ) GmbH

Partner: ITM Isotopen Technologien München SE

Countries or places of project activities: Beijing

15.1 Social Challenge

Malignant tumour has been the leading cause of death among disease-related deaths for a decade in China, and the incidence and mortality rate of all types of cancers in China are both higher than the worldwide average level (National Bureau of Statistics of China, 2020). According to the National Cancer Center, 7.5 people were diagnosed with cancer every minute, and there were 2.338 million cancer deaths in 2015 (National Cancer Centre, 2019). In 2018, the three leading forms of cancer among new cases were lung (18.1%), colorectum (12.2%) and stomach (10.6%) (International Agency for Research on Cancer, 2018).

S. Porsche (✉) · S. Cui
Deutsche Gesellschaft Für Internationale Zusammenarbeit (GIZ) GmbH, Bonn, Germany
e-mail: sabine.porsche@giz.de

© The Author(s) 2025   177
Deutsche Gesellschaft für Internationale Zusammenarbeit (GIZ) GmbH et al., (eds.),
Sustainable Aging, https://doi.org/10.1007/978-3-662-69139-7_15

Since 2012, a series of policy measures were taken to reduce the numbers of cancer patients and the burden on the health system in cancer, such as the Healthy China Initiative 2019–2030 (Department of Planning and Information, National Health Commission, 2019) and the Health China Action—Implementation Plan for Cancer Prevention and Treatment 2019–2022 (Bureau of Disease Prevention and Control, National Health Commission, 2019). The Chinese government has started initiating massive screening programmes, including the introduction of innovative methods for preventive screening, early diagnosis and treatment for example based on nuclear medicine. Precision Oncology or Targeted Radionuclide Therapy is one effective method based on nuclear medicine from Germany. However, the application of this method requires well-trained medical professionals and standardised procedures which are currently not in place in China.

15.2 Project Approach

To improve the treatment for the rising number of cancer patients in China, the German company ITM Isotopen Technologien München SE entered a development partnership under the develoPPP.de programme with GIZ, which operates on behalf of the German Federal Ministry for Economic Cooperation and Development (BMZ). ITM produces radioisotopes which can be used in Targeted Radionuclide Therapies for diagnosis and treatment of cancer.

Goal of the project is to provide online trainings for medical professionals to apply Precision Oncology. The development of curricula as well as training material and the establishment of an online-learning platform are in the focus. The online modules offered for five different groups of professionals involved in the application. Professionals that participated successfully received an international certificate. Until end of August 2023, 257 users from 109 hospitals in China have been certified by the project in their respective pillars, 340 certificates in total have been sent out to those users.

Besides the trainings, a multi-stakeholder political dialogue between German and Chinese ministries, government agencies as well as the industry is established to foster the standardisation of procedures applied in Precision Oncology in China. Awareness raising activities offered for patients as well as the public to inform about this method.

15.3 Value Added

The project is aligned with Chinese national health strategies which also contribute to fulfil the Sustainable Development Goals (SDGs). The project contributes to this international agenda and the national strategies by promoting healthy lives and well-being as well as increasing the recruitment, development, training and retention of the health workforce in developing countries (SDG 3 and 3c, UN DESA, 2015). The goals of Healthy China Initiative are to curb the uptrend of cancer and to keep the overall 5-year survival rate of cancer above 43.3 per cent until 2022, and above 46.6 per cent until 2030 (Department of Planning and Information, National Health Commission, 2019). The project contributes to these targets by introducing Precision Oncology to China, establishing standardised diagnostic and treatment methods for its application and providing training to Chinese nuclear medicine professionals on its application in cancer treatment.

15.4 Ideas for the Future

With the kick-off of the online trainings in China a Sino-German community for Precision Oncology will be gradually established ensuring the dissemination and application of this method throughout China. With the rising number of hospitals with standardised treatment in China, the competences and trainings can be shared with countries in Central- and Southeast-Asia and other continents.

References

Bureau of Disease Prevention and Control, National Health Commission. (2019). Notice on publication of the healthy China action, implementation plan for cancer prevention and treatment (2019–2022). Retrieved from: http://www.nhc.gov.cn/jkj/s5878/201909/2cb5dfb5d4f84f8881 897e232b376b60.shtml [06.11.2020].
Department of Planning and Information, National Health Commission. (2019). Healthy China initiative 2019–2030. Retrieved from: http://www.nhc.gov.cn/guihuaxxs/s3585u/201907/e92 75fb95d5b4295be8308415d4cd1b2.shtml [06.11.2020].
International Agency for Research on Cancer, WHO. (2018). Number of new cases in 2018, both sexes, all ages. Retrieved from: https://gco.iarc.fr/today/data/factsheets/populations/160-china-fact-sheets.pdf [06.11.2020].
National Bureau of Statistics of China. (2020). Annual data, statistical database. Retrieved from: http://www.stats.gov.cn/english/Statisticaldata/AnnualData/ [06.11.2020].
National Cancer Centre. (2019). National cancer report. Retrieved from: https://www.cn-healthcare.com/article/20190623/content-520594.html. [06.11.2020].
State Council of the People's Republic of China. (2017). Notice on publication of the medium- and long-term plan of the prevention and treatment of chronic diseases of China (2017–2025). Retrieved from: http://www.gov.cn/zhengce/content/2017-02/14/content_5167886.htm [06.11.2020].
UN DESA, United Nations, Department of Economic and Social Affairs. (2015). The 17 development goals. Retrieved from: https://sdgs.un.org/goals [05.11.2020].

Sabine Porsche graduated in Cultural Anthropology from Philipps-University in Marburg. She joined Deutsche Gesellschaft für Internationale Zusammenarbeit (GIZ) GmbH focusing on the private sector cooperation as well as higher and vocational education. After working in China until 2021 she is now in the Palestinian Territories. From 2007 to 2016 Sabine held different positions at Tongji University in Shanghai, China. She was the German Vice-Director of the Sino-German University of Applied Sciences (CDHAW) and established the DAAD Career Academy.

Shuo Cui graduated from the University of Tasmania, with the master's degree in finance. She joined Deutsche Gesellschaft für Internationale Zusammenarbeit (GIZ) GmbH China in 2018, she has been working for different projects related to climate adaption, sustainable infrastructure standard, elderly care, precision oncology, neonatal care. Currently, she is the advisor for the 'Sino-German Cooperation on Climate Change-NDC Implementation' project.

Part IV
Age-Friendly Cities and Communities

Chapter 16
Adapting Cities to Demographic Challenges of Ageing: A Review of the Political Landscape and Approaches in China and Germany

Marie Peters

16.1 Introduction

China and Germany are facing the same challenge: the ageing of their population. The German population has been ageing since the 1970s with its fertility rates already decreasing since the so-called "baby boomer" generation of the early 1960s (Statistisches Bundesamt, 2012). Ever since the 1980s, this demographic development has been identified by scientists as a societal challenge (Schneider, 2013). China's ageing population is a relatively new and fast-developing phenomenon. In China, the fertility rate has been decreasing roughly since the 1990s and fell from 5 to 1.6 children per woman in 2016, thereby falling below the replacement level (The World Bank, 2018; Whitebrook, 2016). The ageing phenomenon is not only affecting rural areas, but it is increasingly creating consequences in cities. For instance, in 2011, it was found that in OECD countries, 43.2% of all populations above 65 years lived in cities with a higher increase in metropolitan areas (2001 and 2011 23.8% in metropolitan areas and by 18.2% in non-metropolitan areas), whereas more older people live in the urban hinterlands than in the urban centres. In Germany, the percentage of elderly within the total urban population is comparably high (almost 20% compared with the OECD average of 15%) with almost equal shares in the urban centres and the hinterlands (OECD, 2015). Since 2015, more Chinese citizens aged over 60 years reside in urban areas (52%), revoking the past pattern of a primarily ageing rural China (2000 it was still only 34% in urban areas, Yang et al., 2018). The ageing of the population is becoming increasingly apparent in the first-tier cities, representing a future development trend also for smaller cities. In 2023 and 2022, respectively,

M. Peters (✉)
Deutsche Gesellschaft für Internationale Zusammenarbeit (GIZ) GmbH, Bonn, Germany
e-mail: marie.peters@giz.de

© The Author(s) 2025
Deutsche Gesellschaft für Internationale Zusammenarbeit (GIZ) GmbH et al., (eds.),
Sustainable Aging, https://doi.org/10.1007/978-3-662-69139-7_16

the percentage of people over 65 years in Beijing and Shanghai reached 15.9% and 18.7 of the total urban population (Statista, 2023, 2024).

The apparent trend of an ageing urban population in both countries is accompanied by other societal changes, such as dissolving traditional family-support structures and high rates of single child households, leaving many elderly people to care for themselves. This situation is especially the case in urban areas where spatial barriers hinder mobility and anonymity granted by large cities can hinder extra-familial support systems, thereby rendering life challenging for many elderly people. Both countries, China and Germany, have accepted this challenge and have identified it as a major field where action is needed. This chapter aims to give an overview on the institutional landscape dealing with urban adaptation to demographic change in both countries and reviews current national policy endeavours and research efforts to address the challenges.

16.2   Institutional Landscape and Political Approaches in Germany

In Germany, it took 30 years for the political leaders to not only acknowledge the ageing of the population but accept it as a major political challenge that needs to be addressed on highest level. In 2012, the German government adopted their Strategy on Demography. This scheme aims to equip all citizens—regardless of age or living condition with opportunities to develop their potential and skillset. The adjusted Strategy also includes the promotion of societal cohesion and equal living conditions. Intersectoral working groups have started addressing the targets in the same year to give advice on various topics, including self-dependent ageing and the promotion of living quality in cities and counties (Demografieportal des Bundes & der Länder, 2015). The Strategy on Demography is also the guiding document for the political endeavours of the ministries involved in adapting cities to the ageing society. In 2013, Chancellor Angela Merkel identified the ageing of the population as one of the key political priorities and has treated it with the same amount of high urgency as the energy transition, climate change, globalisation and financial crisis (Schneider, 2013).

Since then, several ministries are actively engaged on this issue as they find approaches to adapt cities and other living environments to the ageing society. The Federal Ministry for Family Affairs, Senior Citizens, Women and Youth (BMFSFJ) is responsible for improving the political frameworks for elderly people, for example, the law on elderly care or the law on home care. It initiates research projects to improve the basis for political decision-making, launches pilot projects to support the independent life of the elderly and improve elderly care, it supports European and international cooperation on the national level, and promotes associations for senior citizens in their line of work (BMFSFJ, 2018). Their work is closely interlinked with endeavours of other ministries. The Ministry of Transport, and Digital Infrastructure (BMVI), for instance, engages in projects on accessibility of public services for elderly among others and the Ministry of Health creates political framework conditions for good practices in health promotion, elderly care and health prevention and rehabilitation (BMG, 2018; BMVI, 2016). The Ministry for Education and Research (BMBF) focuses on research projects to develop technical innovations that can support the elderly, nurses as well as family caregivers (BMBF, 2019). However, the focus in these ministries lies primarily on accessibility and elderly health.

16.2.1   Research Endeavours of the German Government

Since 1993, the German government (led by BMFSFJ) publishes an Old Age Report every four years drafted by an independent academic commission. The Report gives support in political discussions and decision-making by providing information on current trends, needs and challenges (Deutsches Zentrum für Altersfragen, 2018). The fifth Old Age Report published in 2006 firstly addressed the topics of social integration, such as family and private networks, engagement and participation in society. It, thereby, acknowledges that many elderly are still active in their early retirement ages and willing to actively contribute towards the wider community. The seventh Old Age Report published in 2015 addressed the role of municipalities in sustainable ageing. This report highlights that living at old age should be independent and self-determined, whilst it further presents that one's quality of life is contingent upon the quality and diversity of their living environment and social integration (BMFSFJ, 2015). This indicates that Germany's elderly policy emphasises both an integrative and sustainable approach to ageing, while it also recognises the central role municipalities can play throughout the whole process. Furthermore, the Report identifies concrete recommendations for action on the local level regarding providing public services, promoting affordable and barrier-free housing, developing technical support systems, ensuring mobility and accessibility, designing social space, and promoting of neighbourhood relations. A key take-away from the report is that successful housing policy for the elderly is an interdisciplinary field that needs multi-stakeholder coordination and inclusive policymaking as the elderly can also benefit from policies not specifically drafted for elderly (BMFSFJ, 2015)—a recommendation that applies not only to the field of housing but the general area of ageing politics. In August 2020, the eight Old Age Report was published highlighting the role of digitalisation in providing support for an independent lifestyle of the elderly. It addresses fields such as housing, mobility, social integration, health, elderly care and social space. Whereas the report was finished before the begin of the COVID19-pandemic, the results are even more relevant according to the BMFSFJ as especially in times of crisis digital solutions provide great support for elderly in keeping up their daily activities (BMFSFJ, 2020).

16.2.2 Piloting

Besides conducting research projects, the German government engages in pilot projects to find solutions for the ageing challenge in German cities. One of these projects is an initiative initiated in the federal state Lower Saxony and in a later stage supported by BMFSFS to pilot the concept of the "modern extended family". The pilot investigated how both young and elderly unrelated residents of a building voluntarily can support each other in daily life as well as live within an open community. The aim was to foster social interaction between the generations and promote self-supportive capabilities among young and old residents. For example, senior participants could provide daily care for resident families' children and in return the families could provide support in daily shopping activities. Potential sponsors of such projects include the municipality, religious institutions, or other associations. In 2005, the programme was initiated on the national level in three stages from 2006 until 2020. Since 2012, the programme follows the four key targets of old age and elderly care, social integration and education, services in proximity and voluntary work. The focus of the second stage, which concluded in 2016, was promoting employability. This aim was achieved by launching multi-generational houses developing services to better reconcile work and family, thereby supporting jobseekers and single parents. Therefore, the pilot not only explored ways for coping with the demographic change of an ageing population but also addressed this phenomenon in a more holistic manner. In June 2018, there were around 540 multi-generational houses nationwide. As part of the scheme, all households received financial support from the national and the municipal government on a yearly basis (BMFSFJ, n.y.; DStGB, 2014).

Another piloting project was implemented by the Federal Centre for Health Education on behalf of the Ministry for Health within the framework of the nationwide programme "Ageing in Balance". Between 2015 and 2016, the Centre organised a competition to showcase the various activities of cities and towns in regard to mobility and elderly healthcare, social integration and accessibility of housing and neighbourhoods as good practice examples for other towns and cities. Out of almost 100 contributions, 9 projects from cities, municipalities and counties were selected and awarded with lump-sum payments between 4000 and 10,000 Euros. The contributions showed that the competitors developed services for elderly people based on the demands of the elderly and that more conceptual work is essential to achieve a better interlinkage of the services with overarching, citywide and countywide concepts. Furthermore, most of the services targeted the elderly aged between 65 and 80 years—either still being active or merely mobility-impaired—, thereby excluding younger and older citizens—often care-dependent seniors—from participating in the scheme. The results of the competition showed clearly that voluntary work is key to providing support services for the elderly, not only in their operation but also in the process of demand identification. Therefore, volunteering should be further encouraged. Additionally, it was possible to show that the sustainable implementation of measures and services and the operation with broad effect needs widespread, well-maintained cooperation. Here, networks of municipalities with sports associations and the cooperation with health insurers to ensure financial means for elderly care, for instance, can be named important (Difu, 2016). Whereas the competition has showcased the success factors of elderly support services, including cooperation among various stakeholders and profound demand analysis, the Federal Centre for Health Education and its partner institutions fail to foster the process of knowledge transfer to other cities, towns, municipalities and countries in Germany. Some of the best practice examples from the competition implemented by local actors will be presented in the following section.

In addition, the Ministry of Transport, and Digital Infrastructure (BMVI) was involved in the implementation of pilot projects in the field of city adaptation to the ageing society. It implemented a pilot project programme in 18 regions from 2016 to 2018 on accessibility of public services with a specific focus on the needs of the elderly among other issues (BMVI, 2016). The developed concepts for cooperative provision of public services in a region ("Kooperationsraumkonzept") can be used as good practice by other regions with shrinking and ageing population (Jahnke et al., 2018).

16.2.3 Best Practice Examples

Socio-Spatial Mapping and Analysis for an Age-Friendly Urban Quarter
In one of its urban quarters, the city of Weinheim undertook an analysis of its socio-spatial situation in regards the city's ability to enable elderly people to live an independent life as long as possible. The analysis consisted of the following tasks:

- Population and structure-analysis with mapping;
- Participatory excursions for the urban residents to explore and analyse the urban quarter they are living in regarding the support system for elderly citizens and living environment;
- Network analysis of the relevant stakeholders in the urban quarter;
- Qualitative interviews with elderly people living alone to discover potential barriers and challenges to their more active participation in society.

One of the project's successful outcomes related to the high involvement of residents. The city administration benefited from the knowledge exchange of the residents as this enabled a better understanding of the participants' needs and challenges. The residents' involvement triggered new dynamics and triggered a community spirit as participants offered to volunteer and contribute, such as by donating public benches or offering trainings for the use of rollators on buses. The project was highlighted as a good practice by the Cooperation Network "Equity in Health" and the Federal Centre for Health Education (Brandeis, 2015).

Assistance Service on Bus Rides for Seniors from Seniors
In 2009, the German city of Lippstadt kicked of a project that involved elderly members of the community volunteering to assist mobility impaired senior citizens during their bus journeys throughout the city. On a regular basis on specific weekdays, a total of 20 the above volunteers help elderly people in need with various practical tasks including getting on and off the bus, storing shopping bags, rollators or wheelchairs, etc. Through this assistance, the volunteers provide services which the bus drivers cannot provide with such intensity and enable mobility-impaired elderly to use public transport as a regular means of transportation. The volunteers undertake several types of training before entering the service, such as on first-aid, bus safety, managing conflict and bus schedule information (Difu, 2016; Bundeszentrale für gesundheitliche Aufklärung, 2016). The project also participated in the competition organised by the Federal Centre for Health Education but was not awarded.

Intergenerational Exchange Between Teenagers and Seniors "17/70"
This project was initiated by the agency for voluntary work in the city of Essen within the aim of promoting a strong social foundation whereby all generations support each other. The project strives to encourage teenagers to participate in voluntary work as well as enhance their skillset and foster the intergenerational interaction. The scheme welcomes teenage participants aged 14 years and above, who are then encouraged to spend time with the senior citizens by reading to them, attending events together or jointly participating in group activities. The concept of this intergenerational exchange was developed by social education workers, psychologists and practitioners from elderly care. The young volunteers are trained by professionals and work in elderly care homes for a period of about 10–12 months. Regular supervision and one-on-one discussions with the young volunteers are an integral part of the work. The volunteers are trained in key and methodological competencies, such as acceptance, empathy, self-initiative and teamwork skills, as well as specialist competencies in terms of medical care for elderly and dealing with dementia.

Intergenerational Park
The intergenerational park in the town of Ortenburg in South Germany was designed in 2015 as an integrative park for all age groups. In comparison to other parks, it puts specific emphasis on the needs of elderly people in the community and is supposed to act as an intergenerational meeting place. It is equipped with gaming devices for children as well as sport equipment for the elderly. Interaction between the different age groups is encouraged by specified service offers (Bundeszentrale für gesundheitliche Aufklärung, 2016). The project was a runner-up in the competition organised by the Federal Centre for Health Education.

16.3 Institutional Landscape and Political Approaches in China

Before entering the twenty-first century, institutional elderly care in China was rare and limited to the so-called "Three No's"—people with no children, no income, and no relatives, who were publicly supported welfare recipients (Chen, 1998). Thus, few services and institutions were provided in urban areas and support was mainly provided by the elderly's family networks. Although the times when elderly care was ingrained in Chinese Confucian culture as filial piety has long gone, due to the ageing population and the family support systems dissolving, the discussion about age-friendly environments in Chinese cities started to evolve. In China's 12th Five-Year Plan (which is the major national-level policy document guiding the country's development and directing action on provincial, county and city level for the period of 2011–2015), the topic of the ageing population was first addressed. Consequently, several national policy initiatives have started to develop a supporting system for the elderly in dealing with the challenge of an ageing (urban) population. This policy shift opened a new prospect allowing the government to play a stronger role, whilst it also welcomed private sector involvement in elderly care. The 13th Five-Year Plan has also spurred the strategic shift from elderly issues being a mere family topic to becoming a social issue, focusing on not only health and care aspects but also those associated with living and well-being.

Due to the interdisciplinary nature of the topic, the institutional landscape in China—same as in Germany—is quite diversified, with the Ministry of Civil Affairs (MoCA) and the National Working Commission on Ageing[1] (NWCA) acting as two major forces which address the challenges of an ageing population, whereas other ministries and implementing agencies are also responsible for such tasks (see Table 16.1). In charge of building the political framework for social and administrative affairs, MoCA sets national policies and regulatory guidelines that are to be implemented by provincial-level officials. The NWCA was already established as an advisory and coordinating organisation of the State Council in 1999. However, its gained greater responsibility as the ageing society became increasingly recognised as a major challenge. These days, NWCA initiates extensive research projects, launches pilot schemes and coordinates policy making, planning, developing elderly care services nationwide through a central-to-local work network. The network of the commission and its offices were formed at the provincial, municipal, county and township levels. Not only because of the interdisciplinary nature of the topic, but also due to the greater challenge that the ageing population has posed to the society as well as NWCA, the work of NWCA became more closely interlinked with endeavours of other ministries. The Ministry of Housing and Urban–Rural Development

---

[1] Until 2018, the tasks of the NWCA implemented in its secretariat was undertaken by the MoCA although the NWCA was an independent institution under the State Council. Since the restructuring of the State Council in 2018, the duties of the NCWA were integrated into the National Health Committee (NHC), a national ministry under the State Council responsible for formulating policies and measures to deal with the population ageing and combining elderly care with medical care.

(MoHURD), for instance, takes the lead in the construction of age-friendly communities and cities, with the knowledge support from its Centre of Science and Technology and Industrialisation Development (CSTID). Under the guidance of the Ministry of Housing and Urban–Rural Development (MoHURD), the Centre of Science and Technology and Industrialisation Development (CSTID) has carried out the research work on age-friendly communities and cities. CSTID receives commissions from governmental agencies and other organisations for consultation services and project evaluations, participates in the formulation of technical standards, to lay a strong scientific foundation for the government's decision-making process on city adapting to demographic changes.

Table 16.1 Distribution of responsibilities between ministries and institutions for tasks related to developing age-friendly environments in Chinese cities

Task	Leading responsibility	Implementation responsibility
Development and establishing of elderly service facilities (e.g. general health, care, etc.) in new and existing municipalities and urban quarters		National Development and Reform Commission (NDRC), Ministry of Civil Affairs (MCA), Ministry of Finance (MoF), Ministry of Housing and Urban–rural Development (MoHURD), Ministry of Commerce (MoFCOM), National Health Commission of PRC (NHFPC)
Establishment of sports facilities in public spaces	General Administration for Sport	State Ethnic Affairs Commission (seac), MoHURD, National Working Commission on Ageing (NWCA), Ministry of Culture and Tourism (MCT)
Planning and establishing barrier-free urban environments (public spaces and buildings)	MoHURD	NDRC, Ministry of Industry and Information Technology (MIIT), MCA, MoF, Ministry of Transport (MoT)
Age-friendly renovation of urban quarters and buildings with high living standard (green and safe), energy-saving renovation of elderly service facilities, establishment of forms of living that support inter-generational living, establishment of support institutions for the elderly in urban quarters	MoHURD	NDRC, MCA, MoF, NWCA

Source Own compilation based on NWCA (2017)

16.3.1 Current Policies and Research Endeavours

Not until 2013, the National Working Commission on Ageing (NWCA) first acknowledged the importance of a liveable environment for the elderly and underlined the indispensability of "age equality" during urban construction and management in its *Report of the Development on Ageing Cause.* This report, which was first issued in 2006, is one of the most important policy documents exploring the development process of the ageing population in China (Wu, 2013). However, in contrast to the 2006 Report, the follow-up version serves only as a guideline supported by governmental institutions instead of being a governmental directive, thereby having less impact. A further publication—*the Report of the Development on Liveable Environment for the Elderly*—was issued in 2015 by the Centre for Ageing Science Research (under NWCA) and the Faculty of Architecture at Tsinghua University to present several strategies, such as fostering the development of an age-friendly (public) environment and the promotion of liveable communities for the elderly as well as the establishment of related standards. This Report sought to address the imbalanced developments of elderly care in urban and rural areas as well as the healthy development of an elderly care industry.

These research reports have assisted in the formulation of a sectoral 13th Five-Year Plan (sFYP) on national level on how to address the topic of an ageing society and establish an elderly care system. Serving as a guiding document for all provinces and municipalities, the sFYP released in 2017 identified the directions and the targets to tackle the challenges of the ageing society in China. In keeping with the community-based approach adopted by the government's elderly care policies, CSTID has formulated a series of technical standards on construction, management and development of residential communities for the elderly, which identify concrete instructions for implementation, bridging the gap between policymakers and policy implementers. Apart from ensuring mobility and accessibility of community infrastructure and services, CSTID is also dedicated to incorporating artificial intelligence and environmental sustainability within its adaptation approach. As a result, CSTID keeps abreast of the opportunities and challenges in this progressive era.

Further, under the guidance of MoHURD, CSTID has developed a standard system for age-friendly elderly care facilities. This standard system offers many functions, such as setting accessibility criteria for the buildings, but also suggestions on the design of community spaces, for instance, in regard to the distance to service facilities or the provision of resting spaces in public areas. In his article in this book, Youyang Zhao explains the importance of such a standard for China, granting a detailed insight into the drafting process.

16.3.2 Piloting

The Chinese Ministry of Housing and Urban–rural Development (MoHURD) further improves the relevant technical standards for building by piloting different projects through its sub-institution CSTID. One project currently in its implementation phase is a comprehensive urban renovation project of the Daliushu Community in Haidian District in Beijing, which focuses on adapting the urban environment, public service facilities and buildings (see Fig. 16.1). The renovation of the urban environment includes the barrier-free renovation of sidewalks, public spaces, parking lots, toilets, the inclusion of resting seats and handrails as well as barrier-free signs (i.e. for people with disabilities) in public areas. Public facilities, such as neighbourhood committees, health stations, community centres and so on are also made barrier-free, for instance, through the installation of wheelchair ramps, lifts or elevators and clear signs. The renovation of existing buildings to improve accessibility mainly focusses on retrofitting buildings with elevators as well as renovation inside the apartments.

There were key challenges encountered in the implementation of the renovation. Firstly, one challenge regarded convincing the residents to financially support the renovation of their buildings and their immediate environment. Secondly, in the implementation, several departments were required to grant approval of the renovation which led to further complication and difficulty. Similarly, it also became clear in the implementation process that the needs and demands of various interest groups, like residents or enterprises, needed to be considered. When it came to decision-making, balancing the interests of various parties turned out to be key in the successful implementation. Finally, uncertainties at the policy and regulatory level made the implementation difficult but also showed that the implementation of pilot projects was important to advance existing policies and standards (Zhao, 2024).

Fig. 16.1 Renovation in process in Daliushu Community, Beijing (*Source* Youyang Zhao)

Another pilot project is implemented by the local department of MoHURD in Jiangsu Province. The provincial government has established a fund for the renovation of existing buildings in 2017 with the purpose of promoting the communities to conduct their own age-friendly renovations. In the city of Changzhou several communities have made use of this funding instrument, such as the Heyuan West Village or Huadong Community, which both have a high proportion of elderly people within their population. In these communities, the renovations not only include technical aspects, such as elevators or ramps but the renovations also include the establishment of social services, such as leisure facilities, cultural activity centres, or daily conversation- and housekeeping-services provided by volunteers. In Heyuan West Village, the renovation took place with participation of representatives of the community who provided suggestions and communicated relevant needs.

Besides these pilot projects implemented by the national government and its sub-institutions, local actors in the cities initiate various initiatives to adapt their environments for elderly people. Some of these will be presented in the following section.

16.3.3   Best Practice Examples

A new mechanism of community-based elderly care, Shahekou District in Dalian.

As a traditional industrial base in the Northeast of China, Dalian has faced the problems of its ageing population and high unemployment rate due to the reform of the economic system. Therefore, Dalian needed solutions to remedy the above-mentioned problems. In the provision of community-based elderly care, Shahekou District in Dalian has found its own path by taking innovative measures. In 2002, the District had 182 elderly people who had insufficient daily care assistance as they either had no children or their children did not live nearby. In addition, there were almost 300 unemployed women in the same community. The neighbourhood committee decided to match the care need from the elderly with the employment demand from the unemployed. As a start, the committee organised training programmes for the unemployed women and prepared them with professional nursing skills. Only after being retrained as qualified caretakers, the women could provide the elderly with nursing care and received payment from the committee and a local charity organisation. Through this model, Dalian's seniors not only received support in their daily lives, but a network of community-based social interactions and support could be established, partly substituting the loosening family support structures. This innovative approach immediately caught the attention of the municipality and ever since then this model has been popularised and developed into a more mature mechanism of community-based elderly care.

16.4   Old Fellow Project, Shanghai

Initiated and supported by the Shanghai government in 2012, the long-term Old
Fellow Project organises voluntary activities among elderly people. This sees recently
retired people volunteering to offer care and help to a more vulnerable age group.
On the one hand, most of the recently retired people are in a good health condition
and are also enthusiastic about the voluntary activities; on the other hand, the generation gap among them is much smaller compared to that of the young support staff,
which suggests that the elderly volunteers can understand the needs of their peers
better. There is a great variety of the voluntary activities, for example, telephone
conversations, home visits and so on. By the end of 2014, there were 26 institutions
participating in the Old Fellow Project across 17 districts and counties in the city.
More than 150 thousand elderly people were benefiting from the project and around
30 thousand old volunteers have participated as support givers.

16.5   A University for 70-Somethings

Rudong County is just a sleepy backwater town outside of Shanghai. However, the
town was thrust into the public glare due to its radical and yet simple answer to its
rapidly ageing population: the University of the Aged. As an early testing ground for
the one-child policy, Rudong has to face the ageing crisis way earlier than the rest of
the country, with 30% of its population aged 60 and older in 2017. In the University
of the Aged, pensioners can take part in numerous classes ranging from playing
the flute to learning calligraphy, for just RMB 80 (EUR 11) per term. According to
The Guardian, since the establishment of the university, many local elderly people
are looking at the future with more hope again and feel rejuvenated. Although the
university cannot change the demographics, it is an innovative adaptation to the
demographic changes (Phillips, 2017).

Housing renovation for elderly households.

In 2018, in line with the 13th Five-Year Plan of Ageing Development in Shanghai
(a province-level policy directive following the national 13th Five-Year Plan (2016–
2020)) the local Department of Civil Affairs and the local office of NWCA of
Shanghai Municipality invested RMB 20 million (EUR 2.8 million) in housing renovation projects. This move aimed to increase convenience in elderly's life in terms of
safety, accessibility and hygiene, with kitchens and bathrooms being the main target
of the renovations. Key criteria for qualifying in the scheme included the housing
condition as well as the personal situation of the elderly candidate.

16.6 Conclusion

As urban populations across the world continue to grow, cities will increasingly benefit from a diversifying society due to changing lifestyle patterns, higher life expectancies and improving health statuses in old age. However, cities will also grapple with the challenge of diversifying needs and demands of the ageing population, thus will become centres where flexibility, new ideas and innovation are urgently needed. By improving access to both urban infrastructure and society, cities can ensure people remain active in their work, family and community as they age. Such an approach would guarantee that cities could reap the benefits of the elderly's experience and knowledge as they are actively participating in society. Ageing societies also present opportunities for innovation—and, therefore, development.

This chapter presents that both countries, Germany and China, have started to tackle the challenges their cities are facing by not only providing adequate living environments for the elderly but also creating inclusive societies and equal opportunities for all age groups. On the one hand, the national governments have started to expand responsibilities for existing institutions to address the need for technical and societal adaptation of cities to their ageing population or have established new institutions. This process has created a diverse landscape of institutions which conduct research and implement pilot projects in various fields related to urbanisation and the urban society. Whilst it is apparent that a challenge of such a complex nature can only be tackled by actors with distinct expertise, relying on multi-sectoral management also bears the risk of fragmenting critical work if the approach is poorly integrated.

In China's case, for instance, several bodies take the lead in similar fields at the same time. The result is a policy landscape that is not imperative and lacks consistency and clear implementation. For example, at present, China's pension service policy is mostly limited to the level of ministerial regulations and normative documents. Most of them are formulated and circulated in the forms of "notices", "opinions", "decisions", which have a guiding function but no binding legal character.

A similar picture is painted by the piloting approaches applied in the two countries. In Germany, for instance, a well-structured piloting process was implemented on the national level and even received financial support over the duration of several years (e.g. granting awards to projects on multigenerational housing or active ageing). However, vast potential lies in upscaling the derived knowledge and expertise. Yet a clear strategy needs to be developed in this area, which presents a key task for upcoming legislative periods. On a similar note, it remains extremely important to strategically involve multiple actors—not only actors on the ministerial level but also those on the city-level such as social associations, real estate companies, or medical service centres.

On the other hand, the ageing challenge has created a landscape of innovative local approaches, which arisen from attempts to meet the needs and demands of the elderly—instead of waiting lengthy political processes to solve the daily challenges brought on by the ageing society. These approaches relate to the needed innovation that builds on the knowledge and experiences of elderly people and are vital for the successful adaptation of cities to make them more liveable and socially equal places. Thus, an integrative approach to age-friendly cities lies in the strategic political and institutional action that supports and works in harmony with local initiatives on the ground.

Acknowledgements This research was conducted in the framework of the Sino-German Urbanisation Partnership funded by the German Federal Ministry of the Environment, Nature Conservation and Nuclear Safety (BMU) in cooperation with the Ministry of Housing and Urban-Rural Development of the People's Republic of China (MoHURD) and implemented by the Deutsche Gesellschaft für Internationale Zusammenarbeit (GIZ) GmbH. The author´s thanks go to Hehui Zhang and Meng Lin for their research support as well as to the implementation partner of the Partnership, the Centre of Science and Technology and Industrialisation Development (CSTID), for their technical input.

References

BMBF, Ministry for Education and Research. (2019). Pflege durch Forschung erleichtern. Retrieved from: https://www.bmbf.de/de/pflege-erleichtern-5479.html [04.04.2019].

BMFSFJ, Bundesministerium für Familie, Senioren, Frauen und Jugend. (2018). Retrieved from: https://www.bmfsfj.de/ [04.10.2018].

BMFSFJ, Bundesministerium für Familie, Senioren, Frauen und Jugend. (2015). Siebter Altenbericht. Sorge und Mitverantwortung in der Kommune – Aufbau und Sicherung zukunftsfähiger Gemeinschaften und Stellungnahme der Bundesregierung. Berlin. Retrieved from: https://www.siebter-altenbericht.de/fileadmin/altenbericht/pdf/Der_Siebte_Altenbericht. pdf [21.09.2018].

BMFSFJ, Bundesministerium für Familie, Senioren, Frauen und Jugend. (2020). Digitalisierung bietet großes Potenzial für ältere Menschen. (12.08.2020). Retrieved from: https://www.bmf sfj.de/bmfsfj/aktuelles/presse/pressemitteilungen/digitalisierung-bietet-grosses-potenzial-fuer-aeltere-menschen/159744 [21.09.2020].

BMFSFJ, Bundesministerium für Familie, Senioren, Frauen und Jugend. (n.y.). Mehrgenerationenhaus. Wir leben zusammen. Retrieved from: https://www.mehrgenerationenhaeuser.de [18.10.2018].

BMG, Bundesministerium für Gesundheit. (2018). Retrieved from: https://www.bundesgesundhei tsministerium.de/ [04.10.2018].

BMVI, Bundesministerium für Verkehr und digitale Infrastruktur. (2016). Modellvorhaben „Langfristige Sicherung von Versorgung und Mobilität in ländlichen Räumen". Retrieved from: http://www.modellvorhaben-versorgung-mobilitaet.de/veranstaltungen/hintergrundinfo rmation-zum-modellvorhaben/ [03.11.2020].

Brandeis, B. (2015). Auf dem Weg zu einer alternsfreundlichen Kommune am Beispiel der Stadtteilanalyse Weinheim-West. Geschäftsstelle der Kommunalen Gesundheitskonferenz Rhein-Neckar-Kreis und Heidelberg, Heidelberg. Retrieved from: https://www.gesundheitskonferenz-rnk-hd.de/images/WeinheimW_Bericht_web_low.pdf [12.10.2018].

Bundeszentrale für gesundheitliche Aufklärung. (2016). Gesund und aktiv älter werden. Projektplattform. Retrieved from: https://www.gesund-aktiv-aelter-werden.de/projektdaten bank/ [05.10.2018].

Chen, S. (1998). *Social policy of the economic state and community care in Chinese culture: Ageing, family, urban change, and the socialist welfare pluralism.* Ashgate.

Demografieportal des Bundes und der Länder. (2015). Demografiestrategie der Bundesregierung. Retrieved from: http://www.demografie-portal.de/SharedDocs/Informieren/DE/BerichteKonz epte/Bund/Demografiestrategie.html [04.10.2018].

Deutsches Zentrum für Altersfragen. (2018). Geschäftsstelle Altersberichte. Retrieved from: https://www.dza.de/politikberatung/geschaeftsstelle-altenbericht.html [21.09.2018].

Difu, Deutsches Institut für Urbanistik. (2016). Bundeswettbewerb Gesund älter werden in der Kommune – bewegt und mobil. Dokumentation. Berlin. Retrieved from: https://wettbewerb-ael ter-werden-in-balance.de/preistraeger.html [04.10.2018].

DStGB, Deutscher Städte- und Gemeindebund und BMFSFJ, Bundesministerium für Familie, Senioren, Frauen und Jugend. (Eds.) (2014). Kommunale Impulse generationenübergreifender Arbeit. Hintergründe und Einblicke aus dem Aktions programme Mehrgenerationenhäuser. DStGB Dokumentation No. 129. Berlin. Retrieved from: https://www.bmfsfj.de/blob/77460/86f2f0bacf1777d1beff815950aa29e5/dstgb-dokumentation-mehrgenerationenhaeuser-data.pdf [21.09.2018].

Jahnke, M., Albrecht, M., Kis, A., & Lueder, B.-T. (2018). Modelvorhaben Langfristige Sicherung von Versorgung und Mobilität in ländlichen Räumen. Abschlussbericht. Retrieved from: https://www.schleswig-flensburg.de/media/custom/2120_2801_1.PDF?154 8942263#page=60&zoom=100,129,732 [03.11.2020].

NWCA, National Working Commission on Ageing. (2017). Opinions of the national committee on ageing on the implementation of the "13th Five-Year Plan for the development of the aged career and the construction of the aged system" (15.03.2017). Retrieved from: http://www.gov. cn/zhengce/2017-03/15/content_5177770.htm [16.03.2018].

OECD, Organisation for Economic Cooperation and Development. (2015). *Ageing in cities*. OECD Publishing. https://doi.org/10.1787/9789264231160-en

Phillips, T. (2017). China's answer to its ageing crisis? A University for 70 somethings. *The Guardian, Guardian News and Media*. (24.02.2017). Retrieved from: www.thegua rdian.com/world/2017/feb/24/grey-wall-china-rudong-town-frontline-looming-ageing-crisis [24.10.2018].

Schneider, N. (2013). Die demographische Entwicklung in Deutschland und Europa. Trend Ursachen und gesellschaftliche Gestaltungsmöglichkeiten. Vortrag beim KIT Colloquium Fundamentale. (14.11.2013). Retrieved from: https://www.youtube.com/watch?v=wzXHBB vXLvU [11.10.2018].

Statistisches Bundesamt. (2012). Geburten in Deutschland, Wiesbaden. Retrieved from: https://www.destatis.de/DE/Publikationen/Thematisch/Bevoelkerung/Bevoelkerungsbewegung/Bro schuereGeburtenDeutschland0120007129004.pdf?__blob=publicationFile [20.09.2018].

Statista. (2023). Age distribution of the population in Shanghai, China 2015–2022. Retrieved from: https://www.statista.com/statistics/1130402/china-shanghai-population-distribution-by-broad-age-group/, May 24, 2024

Statista. (2024). Population distribution in Beijing, China 2013–2023, by broad age group. Retrieved from: https://www.statista.com/statistics/1135042/china-beijing-population-distribut ion-by-broad-age-group/, May 24, 2024

The World Bank (2018). Fertility rate China. Retrieved from: https://data.worldbank.org/indicator/SP.DYN.TFRT.IN?locations=CN [20.09.2018].

Whitebrook, A. (2016). *Demographic changes in China to 2030*. Strategic analysis paper, Future Directions International Pty Ltd., Dalkeith. Retrieved from: http://www.futuredirections.org.au/wp-content/uploads/2016/11/Demographic-Changes-in-China-to-2030.pdf [20.09.2018].

Wu, Y. (2013). *China report of the development on ageing cause*. Social Sciences Academic Press.

Yang, J., Siri, J., Remais, J., Cheng, Q., Zhang, H., Chan, K., Sun, Z., Zhao, Y., Cong, N., Li, X., Zhang, W., Bai, Y, Bi, J., Cai, W., Chan, E., Chen, W., Fan, W., Fu, H., He, J., Huang, H., Ji, J., Jia, P,, Jiang, X., Kwan, M.-P., Li, T., Li, X., Liang, S., Liang, X., Liang, L., Liu, Q., Lu, Y., Luo, Y., Ma, M., Schwartlaender, B., Shen, Z., Shi, P., Su, J., Wu, T., Yang, C., Yin, Y., Zhang, Q., Zhang, Y., Zhang, Y., Xu, B., & Gong, P. (2018). The Tsinghua–Lancet commission on healthy cities in China: Unlocking the power of cities for a healthy China. *The Lancet Commissions. Lancet, 391*, 2140–2148. https://doi.org/10.1016/S0140-6736(18)30486-0

Zhao, Y. (2024). Statistical indicators for the development of China's elderly care facilities. In GmbH et al. (Eds.), *Sustainable aging* (pp. 207–214). Springer.

Marie Peters is an urbanisation expert with more than ten years of work experience in East Asia, Southeast Asia and Europe. She trained as an urban geographer at the University of Cologne, Germany, with a focus on sustainable urbanisation, climate-risk resilience, and urban planning & development. She holds a Ph.D. in Geography from the University of Cologne. Her Ph.D. research covered Chinese inter-city competition for talents in the mega-urban region of the Pearl River Delta, China. She has also published on international migration and climate adaptation as well as on the dynamics of industrial upgrading. Her general academic interests include the role of migration within urban transformation and climate adaptation. Marie joined Deutsche Gesellschaft für Internationale Zusammenarbeit (GIZ) GmbH in 2015, where she has been working as a policy advisor for different projects related to urbansiation and transport in Germany and China. She is currently based in Berlin and serves as advisor to the German Council for Sustainabile Development.

Chapter 17
Research on Key Technology for Construction Standards for Retirement Villages—Experiences of Elderly Care Tourism in Developed Countries and China's Ageing Population

Nailin Lou and Youyang Zhao

17.1 New Demands and New Trends Brought on by China's Ageing Population

Since 1999, China's over-60s have made up 10% of the country's entire population. China has officially become an ageing society. Based on the results of the country's 6th census, conducted on the 1st November 2010, China's senior citizens over the age of 60 reached 178 million, 13.26% of the entire population; of these, over 65-year-olds totalled 119 million, or 8.87% of the population, making China the only country to have more than 100 million citizens over the age of 65. Up until the end of 2013, China's elderly population exceeded 200 million, making up one fifth of the world's total elderly population, and half of Asia's. And growth continues, reaching 250 million 60-plus-year-olds in China at the end of 2019, a total of 18.1% of the country's population.

The "Annual Report on Fiscal Policy of China 2010/2011" released by the Institute of Finance and Trade Economics of the Chinese Academy of Social Sciences (CASS) further indicates that the following 20 years will be the period of fastest growth of the elderly population in China. In 2035, the country's elderly population is predicted to reach around 418 million, with the number of over 60-year-olds accounting for 28.7% of the total population, and the old-age dependency ratio forecasted to be 0.28. By 2050, the elderly population in China is predicted to reach 483 million, with the over-60s accounting for 34.1% of the total population, and the dependency ratio set to be 0.46 (see Fig. 17.1). China's transition to an "aged" society is inevitable.

N. Lou · Y. Zhao (✉)
Science, Technology and Industrialisation Development Centre of the Ministry of Housing and Urban-Rural Development, Beijing, China
e-mail: 89215031@qq.com

© The Author(s) 2025
Deutsche Gesellschaft für Internationale Zusammenarbeit (GIZ) GmbH et al., (eds.),
Sustainable Aging, https://doi.org/10.1007/978-3-662-69139-7_17

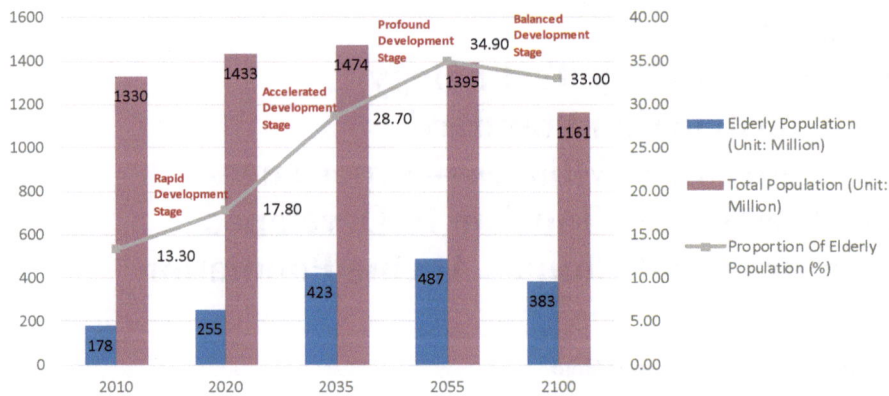

Fig. 17.1 Trends in the number of senior citizens, the total population and the proportion of the elderly population in China (*Source* Own graphic based on Gao, 2010)

The sheer size of the elderly population creates many opportunities and challenges for the development of Chinese society. Home care, community care, institutional care, smart care, migration care[1] and other forms of elderly care have become a topic of general concern across society in recent years. Migration care shows strong growth prospects, having come about with the strong increase in the level of Chinese national economic income and the rise in material needs. Migration care will become a new way to improve quality of life and the course of one's life, especially for China's first baby boomers born in the post-50s and 60s. Figures show that the value ratio of pensions as a source of income for the elderly in China increased from 19.61% in 2000 to 24.12% in 2010. This means that the economic independence of elderly Chinese in their retirement is further enhanced. The constant increase in the proportion of pensions brings with it an equally thriving domestic retirement market. In 2010 alone, 150,000 so-called "migratory bird-like" senior health tourists visited "Longevity Village" in Guangxi province's Bama County. In 2013, 400,000 senior "migratory bird-like" tourists spent the winter in Sanya on Hainan Island. For the most part these tourists come from northern China, known for its harsh winters, flocking to the warmer South like "migratory birds" around October time each year, returning to the North in April the following year.

These distinct features show that the demand for senior care services in China will change greatly in the future. Spiritually and culturally elderly tourists have relatively high demands, they have a strong sense of independence and a certain economic level, giving them more diversified lifestyle choices that require continuous segmentation of the services provided. Whether choosing home care, community care, institutional care or the ever-popular migration care model, the elderly will all participate more willingly and actively. Rising demands for high-quality comfort

---

[1] Migration Care refers to the migrating elderly population that travel away from their hometown to reside in retirement villages, or complexes, for certain periods of the year, returning to their hometown after the travel.

and convenience for aspects such as transport facilities in the care locations, the environment and location of the buildings, intelligent equipment and facilities, the standards of service etc., will result in a sustained increase in the demand for barrier-free access, a boost in safety, bathing and washroom amenities, daily necessities, mobility and walking, professional water waste drainage equipment, and other products including those related to hearing and communication. With this backdrop, the demands for construction of migration care villages should be adjusted to fit in with elderly people's changing lifestyles, to satisfy their migratory life needs in terms of safety, convenience, practicality and comfort.

17.2 Experiences and Insights into Foreign Travel for Elderly Care

The ageing of society was experienced earlier in some developed countries across Europe and America. There is a rich content of study on elderly tourism from these nations, revealing the motivation of elderly tourists. After analysing the living conditions of a sample of migration care seniors in Texas, Guinn (1980) finds that travel motivations for older people include socialising, homesickness, learning, rest and relaxation, exercising and seeking adventure; Anderson and Langmeyer (1982) reveal in their study that people over the age of 50 tend to want to visit, relax or rest; Romsa and Blenman (1989) believe that elderly tourists place importance on visiting relatives and friends and to travel healthily; Thomas and Butts (1998) believe that gathering wisdom, autonomy and social interaction are the main motives for silver-haired tourism; on reviewing previous studies, Fleischer and Pizam (2002) believe that the most common reasons for older people to travel are rest, relaxation, socialising, exercising, learning, homesickness, and seeking adventure; meanwhile Horneman et al. (2002) show that the most frequent motivations as described in previous studies are education and learning, rest and relaxation, physical exercise and fitness, and visiting relatives and friends. Motivation for elderly tourism differs from that of young and middle-aged tourist groups who seek new and different things (positive psychology), or wish to escape from the stress of reality (negative psychology); motivations for elderly tourism are more weighted towards peace, homesickness, family visits, rest and relaxation.

Analysing the reasons behind elderly tourism in other countries and what it entails, can cast further light on China's own migration care market. The emergence and vigorous development of the migration care model firstly depends on the good physical condition of the elderly and rising income levels; secondly, a comfortable climate and pretty natural cultural landscapes are important factors that encourage the elderly to choose migration care; thirdly, excellent public service facilities and medical service systems are important guarantees for living conditions in migration care; fourthly, a diversified migration care model which can pave the way for future development, such as customised travel routes, campervan outings, rehabilitation tourism projects based on improved medical facilities, and so on. The purpose of migration care, the method of travel and the travel process will reveal more distinctive and diverse features in the future.

17.3   The Development of China's Migration Care—Problems and Characteristics

By studying the background of China's ageing population, analysing the experience of elderly travel care abroad, and combining the current situation of the development of elderly tourism in China, the characteristics and problems of China's elderly migration care can be mainly reflected as follows.

17.3.1   Society Lacks Awareness to Promote the Prosperous Development of the Migration Care Market

At present, China's migration care model is not understood or accepted by the whole of society, and there are many misunderstandings that need clarifying. For example, based on traditional Chinese concepts and economic influences, home-based care is certainly the preferred model for most families, but the "4 + 2 + 1"[2] family structure weakens the function of home-based care, and children's care for the elderly falls short. At the same time, in line with elderly peoples' rising income levels and growing self-consciousness, there are many urgent needs for migration care. Many systems in place limit the development of the migration care model; for example, the household registration system is a significant obstacle to elderly tourism in China. Retired seniors are registered with local community management offices, making it difficult for them to benefit from various preferential treatments and welfare offerings at the destination, thus reducing their satisfaction with life and security. In addition, nationwide medical and social insurance have not yet been integrated, which is of concern to the elderly and may over burden public services at the destination. Looking from an international perspective, though China has diverse natural landscapes and rich tourism resources, it does not have special visa arrangements or marketing plans for foreign elderly travellers. The development of inbound elderly care tourism lags that of Malaysia, Thailand and other countries.

---

[2] The typical family structure in China consists of four older people (paternal and maternal grandparents), two parents, and only one child.

17.3.2 Infrastructure for Public Service Facilities and Care Service Facilities is Lacking

Regarding public service and elderly care facilities, there is a lack of suitable barrier-free amenities for the elderly to travel and rest. China's construction of barrier-free environments for the elderly started late and from a low starting point. Comparing the situation with developed countries and regions, there is less investment in China, and there are still gaps in areas including concept, technical standards, monitoring and security systems, systematic and refined design. In many established public environments, barrier-free facilities are not ideal, and complete barrier-free access has not been realised, resulting in the inability to satisfy the needs of users to participate in social and cultural activities on an equal footing.

The barrier-free facilities that have been built are incomplete and not fit for purpose. The overall system is relatively lacking in order and functionality, the accessibility facilities are not universal, the utilisation rate is not high, to the point that even obstacles and injuries are caused to others. As an example, many pedestrian crossings in urban areas have tactile paving (blind sidewalks) built for the blind, but many affect the use of wheelchairs or other pedestrians due to their bad installation.

17.3.3 Existing Programmes for Elderly Migration Care Do not Satisfy the Actual Daily Needs of Elderly Travellers

Elderly migration care is a fusion between so called "migratory-bird" care and "holiday-style" care, and is a way of benefitting the physical and mental health of the elderly to lead an active retirement. It entails migrating to several places during different seasons, usually staying in the same place for about six months or more, travelling while retired and calmly taking it all in to achieve a healthy life and expand one's horizons. A stay from half a month to half a year determines the distinct features of the migration care programme, which should satisfy demands for housing, hotel services as well as security measures for the elderly care institutions. However, China's current migration care programmes are unable to simultaneously cater to the functional requirements of these three key service areas.

17.3.4 Research on Key Technology to Evaluate the Standards of China's Migration Care Villages

It is fundamental to carry out research on the construction standards of migration care villages to actively guide the healthy and orderly development of China's migration care industry, as well as to improve the requirements for support facilities in retirement villages and increase service levels.

The construction standards plan for retirement villages can be divided into five components for evaluation of the overall development; from the surrounding environment and village amenities, the site location, architecture and equipment design, through to smart systems and operational services. The plan reflects the overall quality of the village through a comprehensive assessment of five components and a three-tier classification, to provide consumers with a choice. The purpose is to guide the construction of the village and improve the quality of service and its operation, whilst catering to the choices of different consumer groups.

In terms of the surrounding environment and amenities, the main requirements focus on the environment and amenities on site where the village is located. Environmental aspects include meteorological conditions, natural environmental conditions, cultural landscapes, leisure and places for sightseeing; amenities include transportation, medical and ambulance equipment, commercial facilities, cultural and recreational facilities.

In terms of the site location, requirements centre mainly on the surrounding natural environment, transportation within the village (including connection to the roads and traffic outside the village), green space and recreation areas. Environmental aspects include sunlight, shade, vista, ventilation and distance between buildings; transportation includes the quality of the roads around the village, road width, vehicle speed, internal traffic control, traffic safety, parking arrangements and emergency assistance; green space and recreation areas include plant survival rate, suitable plant species, sensible installation of arbours, plant diversity, surface area of recreation area, facilities for the elderly to rest, venues for night activities, safety and public restrooms.

Regarding architectural design and equipment, requirements focus on the content of three main areas: architectural design, elderly design and construction equipment. Architectural design aspects mainly include construction content, architectural form, spatial layout, circulation requirements, indoor environment; elderly design mainly includes residential buildings, public service facilities, residential facilities, recreational facilities, health care facilities, and safety; construction equipment mainly includes facilities for daily living, water supply and drainage, HVAC, air conditioning, and electrics.

In terms of intelligent systems, there are five main requirements: village services, health services, public safety, smart living, and infrastructure. Village services include systems for smart cards, a call centre, information release, intelligent reservations, environmental monitoring, audio and video multimedia, housekeeping, and supermarket services; health services include health testing services, electronic health records, health assistants, smart care, private doctors, smart mattress rental services, multi-purpose care wheelchair rental services; public safety includes an alarm system, video security monitoring system, access control system, electronic patrol system; smart living includes an indoor information system, smart door lock, indoor intelligent control, indoor equipment control, room management system, artificial intelligence system, smart toilet, and interactive chat; infrastructure includes an information facility system, information application system, automatic fire alarm, computer room engineering, and UPS uninterruptible power supply system.

In terms of operations and services, the main requirements include general management requirements, environmental maintenance, indoor maintenance, operation and maintenance of equipment and amenities, day-to-day services, housekeeping, health care, social entertainment and fitness services, among others.

China's elderly population is currently facing a period of fast growth. The changes in the proportion of the elderly population are large scale, rapid, high peaking, geographically uneven and far reaching. In such a rapidly changing environment, elderly Chinese face a multitude of challenges in their lives. The three most prominent contradictions they face are "ageing before wealth", "ageing before being prepared" and the "rising dependency ratio". It is not difficult to see from the three most prominent contradictions in the life of the elderly that China has prematurely entered the stage of rapid population ageing vis-a-vis society's awareness level and lack of preparation. The social problems caused by this are difficult to eliminate in a short period of time, and the migration care model will be an important part of retirement for the elderly living in China in the future. Actively promoting the construction of environmentally friendly migration care villages will greatly promote the healthy development of China's elderly care service industry.

References

Anderson, B., & Langmeyer, L. (1982). The under-50 and over-50 traveler: A pro-file of similarities and differences. *Journal of Travel Research, 20*(4), 20–24. https://doi.org/10.1177/004728758 202000405

Fleischer, A., & Pizam, A. (2002). Tourism constraints among Israeli seniors. *Annals of Tourism Research, 29*(1), 106–123. https://doi.org/10.1016/S0160-7383(01)00026-3

Gao, P. Y. (2010). *China's Fiscal and tax reform in the period of 12th Five-Year Plan.* China Financial and Economic Publishing House.

Guinn, R. (1980). Elderly recreational vehicle tourists: Motivations for leisure. *Journal of Travel Research, 19*(1), 9–12. https://doi.org/10.1177/004728758001900102

Romsa, G., & Blenman, M. (1989). Vacation patterns of the elderly German. *Annals of Tourism Research, 16*, 178–188. https://doi.org/10.1016/0160-7383(89)90066-2

Thomas, D. W., & Butts, F. B. (1998). Assessing leisure motivators and satisfaction of international Elderhostel participants. *Journal of Travel & Tourism Marketing, 7*(1), 31–38. https://doi.org/10.1300/J073v07n01_03

Wei, J. H., & Cong, Y. G. (2001). Characteristics and the behaviour of elderly tourists and their implications for tourism development (16), 20–23.

Xie, J. J., & Liu, Y. K. (2012). Study report on elderly community care in the United States and Canada. *Property Management in China, 12*, 29–39.

Zhou, C. P. (2002). Thoughts on the expansion of Zhejiang's Yinfa tourism market. *Jinhua Vocational and Technical College Journal* (2).

Zhou, L. (2006). The development of China's elderly tourism market. Master's thesis of Southeast University 28. Retrieved from: https://cdmd.cnki.com.cn/Article/CDMD-10286-2007031531. htm [04.04.2019].

Nailin Lou is the Assistant Chief Engineer at the Centre of Science and Technology & Industrialisation Development of China's Ministry of Housing and Urban & Rural Development. She also serves as the President of Elderly Service Facilities Branch Association of the China Association for Engineering Construction Standardisation. She is a senior-level engineer, and she has been awarded special allowance by the State Council for being an exceptional expert. Nailin is responsible for the development of "Technology standard of the housing performance evaluation", "Construction standard of the community elderly activity centre" and several other national standards. In addition, she also manages the research project regarding "Key technologies and standard systems for elderly facilities' planning and construction".

Youyang Zhao is currently working for CSTID, which is affiliated of China's Ministry of Housing and Urban–Rural Development. He is also the Secretary-General of the Elderly Service Facilities, which is a branch organisation of the China Association for Engineering Construction Standardisation. As part of this role, he researches topics such as elderly care service facilities as well as retrofitting existing buildings to correspond with the needs of an ageing society. He graduated from Birmingham University in the UK, with a major in Urban and Regional Planning from the Engineering of the Centre of Science Technology.

Chapter 18
Statistical Indicators for the Development of China's Elderly Care Facilities

Youyang Zhao

18.1 Background Research

By the end of 2019, China's elderly population of 60-year-olds and over reached 250 million people, equal to 18.1% of the total population. Among them, there are more than 170 million people aged 65 and over, accounting for 12.6% of the total population. Compared with the end of 2018, the elderly population increased by approximately 4.39 million. China's transition to an "aged" society is inevitable. Every aspect related to elderly people's daily lives will face enormous challenges, among them, how to make elderly care facilities that meet the demands of the growing ageing population is of key concern to the whole of society.

For a long time, the construction of China's residential areas has mainly focused on the functional needs of young people, on the set up of auxiliary facilities, such as kindergartens and schools, while, in contrast, there is a significant shortage of recreational and fitness facilities for the elderly in these residential areas. Problems associated with population ageing include a lack of facilities catered to the elderly, leading to more serious social problems, some of which have been reflected in a series of disputes in many parts of the country caused by so-called "square dancing[1]". Likewise, elderly or disabled people face many problems because of residential environment design. Elderly or disabled people need barrier-free facilities with enhanced circulation, ramps and platforms, elevators and level flooring, unhindered indoor and outdoor access and other access areas. Spatial requirements for the turning radius

---

[1] Square dancing is an exercise routine performed to music in squares, plazas or parks. It is popular with middle-aged and elderly retired women in China. Square dancing has been the subject of considerable controversy in the 2010s in China due to complaints of noise pollution in the evening or morning hours, at times leading to violent disturbances.

Y. Zhao (✉)
Science Technology and Industrialisation Development Centre of the Ministry of Housing and Urban-Rural Development, Beijing, China
e-mail: 89215031@qq.com

© The Author(s) 2025

Deutsche Gesellschaft für Internationale Zusammenarbeit (GIZ) GmbH et al., (eds.), *Sustainable Aging*, https://doi.org/10.1007/978-3-662-69139-7_18

of wheelchairs in kitchens and bathrooms is often too narrow and the dimension of internal corridors is not suitable for wheelchair passage. Structural design lacks effective layouts and is not convenient for the growing ageing population. The current situation has attracted the attention of the government, universities, research institutes and various bodies involved in construction and development. Design concepts focus on the daily lives of the elderly from different perspectives, such as barrier-free design, renovation requirements, intergenerational relations, latent design, anti-slip requirements, anti-angular design, smart houses, and equipment suitable for the elderly. The scrutiny of elderly care models such as home care, community care, institutional care and smart care is endless. And this leads to another problem, considering China's massive population base, it is very easy to generate a scale effect as well as excessive development. Thus, ensuring the healthy and sustainable development of elderly care facilities is of particular importance.

Urban areas are a complicated combination of economic, social, cultural, historical, and natural environments, and the consequences of demographic changes and the resulting demands for facilities will ultimately have a major impact on the development of a city. There is extensive use of statistics in urban planning; mathematical statistics find the inherent regularity of a phenomenon by observing the frequency of certain phenomena, then making a judgement and predictions with a certain degree of precision. Statistics can provide the most effective key parameters for planning proposals, by summarising the parameters of past patterns and predicting future development trends, thus establishing an effective basis for planning and design. In recent years, statistical methods have contributed significantly to research on urban resilience and urban sustainable development. Research on a statistical indicator system for elderly care facilities is based on the continuous and in-depth study of the various components of China's elderly service industry; this offers a clear overview of the state of construction of the country's elderly care facilities, and provides a scientific and rational technical basis for each decision-making phase of the research work carried out.

18.2 Why the Need for Such Research?

18.2.1 To Assess the Current Situation of China's Elderly Care Facilities

Currently in China, the classification of statistics and planning for elderly care facilities is not sufficiently uniform between different regulatory bodies and regions. Different terminology is used within statistical work even for elderly care facilities in different departments or regions sharing the same functions. By analysing the current situation of elderly care facilities, it is possible to further clarify the scope of statistics and statistical indicators, and to have a clearer understanding of the status of China's elderly care facilities.

18.2.2 To Clarify Future Development Trends of China's Elderly Care Facilities

Along with the constant acceleration of development and construction, besides analysing the types and scale of existing elderly care facilities, it is also important to predict future development trends. Over recent years, by drawing on the experience of developed countries, many new ideas have emerged vis-a-vis the development of China's elderly care facilities, such as market-oriented participation in co-development, comprehensive elderly care facilities, small-scale and multi-functional elderly care facilities, medical-integrated facilities, and "Internet+" smart methods. Whether any of these new ideas should be integrated into China's future development of its elderly care facilities, should be analysed and predicted through appropriate research work.

18.2.3 To Provide Scientific and Rational Planning for Urban Elderly Care Facilities

The final purpose for research on statistical indicators for elderly care facilities is to provide the basis for cities to undertake scientific and rational planning and design programmes. According to research on statistical indicators, cities can integrate data content from various departments in the region, compare and analyse the data of adjacent regions, and make relatively reasonable future predictions on development trends based on existing data collection and analysis, thus making an appropriately robust construction and development plan.

18.3 Principles

Research on statistical indicators for the construction of elderly care facilities should be based on the long-term development of China's elderly service industry, the research should fully capture the world's advanced and mature technology, adapting this technology to China's basic industry and technology system, in line with China's various regulations and relevant construction standards. By conducting in-depth research into the scope of elderly home care, community care, institutional care, rural care and various other types of elderly care facilities, a statistical indicator system for the construction of such facilities can be established—a system that is of reasonable scale, has clear limits and grading, is affordable and feasible (see Fig. 18.1).

Concise and applicable. The statistical indicator system for the construction of elderly care facilities should assess the status of China's elderly care facilities

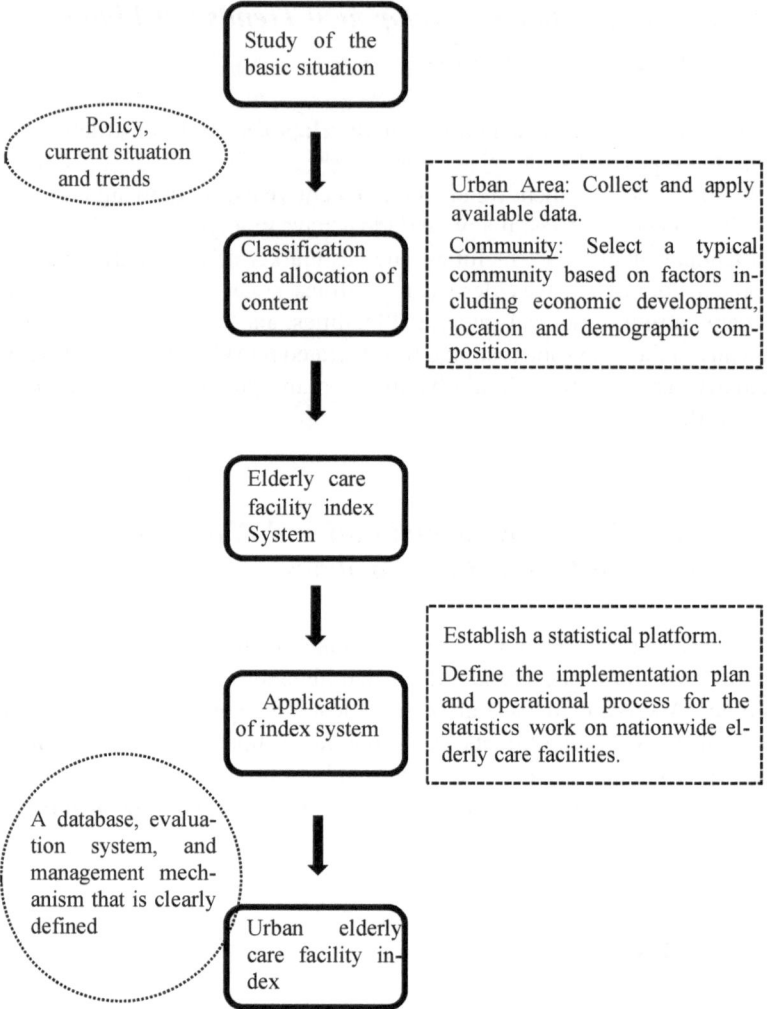

Fig. 18.1 Research flowchart. *Source* own representation based on Zhao

and clearly define research content. On the basis that sufficient research has been conducted, representative and significant indicators are selected to create a statistical indicator system, formulating specific plans for the implementation of the system, as well as explaining the content of the indicators.

Adapted to development. The statistical indicator system for the construction of elderly care facilities should be closely linked to reality and adapted to the development trends of China's ageing population. Studying examples of domestic and foreign long-term cases and thoroughly researching the state of elderly care facilities throughout China, will ensure that the results of the statistical indicator system are realistic, implementable and can be gathered within a certain time frame.

Stable and long-lasting. The statistical indicator system for the construction of elderly care facilities is ultimately an index to evaluate the development of urban elderly care facilities. The composition of the elderly care facility index, and the relationship between a variety of influencing factors, requires a database for support. In addition to first-hand data collected by the initiative, scientific and reasonable linkage data should be used to establish a statistical data platform, thus creating a stable and long-term, adjustable and controllable urban elderly care facility development index.

18.4 Research Highlights

18.4.1 Research the Scope of Elderly Care Facilities and Future Development Trends

After extensively studying the current situation of home care, community care, institutional care, rural care and other types of elderly care facilities in China, and combining existing relevant standards and regulations issued by national and regional authorities, the allocation of China's elderly care facilities is clarified through the actual study of a sample city, according to classification factors such as ownership rights, construction methods, construction status and scope of use. Conducting an in-depth study on the current situation of China's elderly care facilities and referring to elderly care models developed in foreign countries with ageing populations, helps to extrapolate the future development trends of China's elderly care facilities. It further supplements the targeted study into the configuration of China's elderly care facilities, providing fundamental research for the establishment of a classification index system for such facilities.

18.4.2 Statistical Quality and Data Sources for Research into Elderly Care Facilities

At present, the registration and approval authorities for China's elderly care facilities include the civil affairs as well as commerce and industry departments on a local level. The land resource department and the housing and construction departments on a local level are responsible for land supply, planning and construction approval and management of elderly care facilities. Administrative agencies at all levels directly or indirectly participate in the operation, supervision and management of elderly care facilities. Different departments gather different statistical data on elderly care facilities within the limit of their functional authority. Research into statistical quality and data sources is important to establish meaningful indicators and provide a horizontal reference for the establishment of the classification index system of elderly care facilities.

18.4.3   Research into the Applicability of a Classification Index System for Elderly Care Facilities

The statistical indicators and the criteria used to classify elderly care facilities need to align with service facilities on the ground and existing statistical data. The establishment of the indicators should adhere to the following basic principles: accuracy, applicability, adaptability to ongoing development, stability and long effectivity. After the completion of the indicator system, cities with different economic development levels and different degrees of ageing should be selected to evaluate their applicability, thus establishing in one step an elderly care facility classification index system that is practical and feasible.

18.4.4   Research on Demand Analysis, Conceptual Design and Practical Use of Statistical Platforms for Elderly Care Facilities

After completing research on the classification index system of elderly care facilities, and further clarifying the demand analysis and conceptual design of the statistical platforms for these facilities, attention should be paid to the security of the data platform and its ease of use. In the testing phase of the platform, a study will be conducted on the actual usability in the pilot city. This piloting process will ensure data availability for the measurement of indicators of the urban elderly care facilities construction index.

18.4.5   Research on the Rationality of the Factors that Constitute the Index of Urban Elderly Care Facilities

Under the premise that the classification system of elderly care facilities is clearly defined, and the statistical platform is established, further research will be made into the evaluation and management system of urban elderly care facility indicators. Having assessed the current situation of China's elderly care facilities, identified future trends and established a statistical database, the key factors that will form the index of urban elderly care facilities will be determined, and can be investigated rationally. Thus, realising the construction of urban elderly care facilities using an index, and guiding the rational development of China's urban elderly care facilities through intuitive, universal and robust indicators.

Bibliography

Cai, Y. N., & Wen, Z. P. (2017). Climate adaptability planning technology for urban resilience promotion. *Planners, 33*(8), 18–24. Retrieved from: http://planners.com.cn/web_up_file/print_read/2017-8-28913373850_print_read.pdf [24.05.2021].

CNCA, China National Committee on Ageing. (2015). General report on the national strategy for population ageing. *Scientific Research on Ageing, 22*(3), 6–40.

Hou, H. L. (2016). The difference in urban public service supply and its influence on population movement. *Chinese Journal of Population Science, 36*(1), 118–125. Retrieved from: http://www.doc88.com/p-9592338007835.html [24.05.2021].

Li, X. L. (2016, March 23). *Research on the development of our private pension service industry.* Fujian Normal University.

Liang, Q. C., Li, Y., & Shi, Z. G. (2016). Deep causes of square dance disturbing and its governance. *Journal of Beijing Sport University, 39*(1), 26–31.

National Bureau of Statistics of China. (2019). *Statistical communiqué of the People's Republic of China on the 2019 national economic and social development.* Retrieved from: http://www.stats.gov.cn/tjsj/zxfb/202002/t20200228_1728913.html [24.05.2021].

Wei, D. K., Kang, J., et al. (2018). *Standard for design of care facilities for the aged JGJ450-2018.* Retrieved from: http://download.mohurd.gov.cn/bzgg/hybz/JGJ%20450-2018%20%E8%80%81%E5%B9%B4%E4%BA%BA%E7%85%A7%E6%96%99%E8%AE%BE%E6%96%BD%E5%BB%BA%E7%AD%91%E8%AE%BE%E8%AE%A1%E6%A0%87%E5%87%86.pdf [24.05.2021].

Wu, H. (2017). *Research on the development of China's elderly care industry against the backdrop of its ageing population.* Jilin University. CNKI:CDMD:1.1018.007029.

Youyang Zhao is currently working for CSTID, which is affiliated of China's Ministry of Housing and Urban-Rural Development. He is also the Secretary-General of the Elderly Service Facilities, which is a branch organisation of the China Association for Engineering Construction Standardisation. As part of this role, he researches topics such as elderly care service facilities as well as retrofitting existing buildings to correspond with the needs of an ageing society. He graduated from Birmingham University in the UK, with a major in Urban and Regional Planning from the Engineering of the Centre of Science Technology.

Chapter 19
Shaping Inclusive Urban Districts: DGNB-Certification as a Tool for Age-Appropriate Quarters

Christian Eichinger

19.1 Introduction

More and more people live in cities: In 2016, 54.5% of the world's population lived in cities, this number is expected to climb to 60% by 2030 (UN DESA, 2016). At the same time, the world's population is getting older, as populations in both developed and developing economies live longer and healthier lives. Not only emissions, resource- and energy-use are concentrated in cities (World Bank, 2010), but also demographic challenges. A recent report of the OECD outlines that the population share of those older than 65 years of age is expected to climb from 17.8% in 2010 to 25.1% in 2050, with 43.2% of this older population living in cities (OECD, 2015). Therefore, cities take the key role in the discussion of how to achieve a sustainable future that needs to integrate high living standards for all inhabitants, energy-efficiency and limited consumption of resources.

As the real estate sector is very energy- and resource-intensive (UN IEA, 2017) environmental concerns have been a topic since the 1980s. As an answer to very diverse approaches to sustainability, different building certification schemes started to be developed in the 1990s. The goal of these certification programmes was to quantify the effect a specific project had on its environment and make it comparable. Most early certification schemes focused purely on energy-efficiency (e.g. Passivhaus) and were only used to certify single buildings.

In recent years, the focus has shifted from single buildings to the urban scale. Buildings with different functions use energy and resources at different times of the day and can be balanced against each other. District heating and cooling have been introduced in the urban context to optimise energy-use. These interdependencies offer tremendous potential and scale for sustainable solutions that are not possible with a single building.

C. Eichinger (✉)
KSP Jürgen Engel Architekten, Berlin, Germany
e-mail: c.eichinger@ksp-engel.com

© The Author(s) 2025
Deutsche Gesellschaft für Internationale Zusammenarbeit (GIZ) GmbH et al., (eds.),
Sustainable Aging, https://doi.org/10.1007/978-3-662-69139-7_19

This is the reason international building assessment tools such as BREEAM (U.K.) and LEED (U.S.) have developed their own versions of neighbourhood sustainability assessment tools. They also exist in Japan (CASBEE-UD) and Australia (Green Star Communities). Most of these can be categorised as 'spin-off tools', which means that they have been developed from certification schemes focused on single buildings.

In 2007, the "Deutsche Gesellschaft für Nachhaltiges Bauen (DGNB, German Sustainable Building Council[1])" programme was started as a cooperation project between the Federal Ministry for Transport, Building and Urban Development (BMVBS) and members of the planning and construction industry in Germany to originally develop a sustainability standard for new federal and public buildings. From 2009 to 2012 the certification programme was extended to include the DGNB Urban District certification as a neighbourhood assessment tool.

In principle, its understanding of sustainability is based on the "Leitbild Nachhaltigkeit" (Sustainability Guidelines) formulated by the German Parliament (Bundestag). The guidelines describe sustainability as a three-pillared concept (three bottom lines) with the pillars being: ecological sustainability, economic sustainability and socio-cultural sustainability. This broader, more contextualised understanding of sustainability leads to the consideration of the afore-mentioned demographic challenges in the certification programme, mostly through the concept of inclusion.

Inclusion aims to integrate the needs and requirements of all users and inhabitants in an urban district and age-appropriate design is part of this holistic approach. Principally, the focus on the elderly does not mean that other groups, for example families and children, are considered less important. But as demographic changes occur in our societies, the urban framework needs to adapt to the new challenges.

Social inclusion has also found its way into the Sustainable Development Goals of the United Nations that are listed as goals of the DGNB certification programme as well, for example in goal 11: Make cities inclusive, safe, resilient and sustainable.

Studies have shown that the fear of loss of independence and of self-determination is the basis for people's fear of old age (BMFSFJ, 2000: 241; Brech, 2004: 39; Heye & Wezemael, 2007; Mollenkopf et al., 2004: 345). This fear explicitly includes the loss of one's own residence and familiar surroundings in old age. Inclusive urban districts therefore have to allow for an independent and self-determined life of their inhabitants, the ability to stay in their own residence and their familiar environment with its grown social connections. This is not only important for people's personal identity, but it is a determining factor for happiness in old age as well.

This article will first introduce the DGNB system and point out some of the requirements that shape inclusive urban design. It will then consider the implementation and adaptation to the circumstances China.

---

[1] For further information see: www.dgnb.de.

19.2 DGNB Certification for Urban Districts

DGNB certification for urban districts awards the degree of sustainability of a neigh-bourhood with a platinum, gold or silver rating. Such visualisation of area-based sustainability provides stakeholders with a simple tool to gather information about an area. For municipalities and developers, it can be a way to gain an overview of various sustainability issues. At the same time, it can make an area more attractive for future investments, as it shows that sustainability risks have been addressed. The certification aims at making sustainability concepts explicit and allows for consistent benchmarking across districts and cities, putting an end to mere "greenwashing".

The DGNB System has developed usage profiles for district certification to account for the specific requirements of Urban Districts, Business Districts, Industrial Areas, Event Areas, Resorts and Vertical Cities. Going further, this text is going to take the Urban District classification as its basis to show the usage for age-appropriate quarters.

DGNB certification rankings are based on the entire life cycle of a development and offer transparency and comparability between projects. Through the life cycle cost assessment of a development (costs of construction, operation and maintenance costs directly related to owning or using the project), a thorough cost–benefit analysis is integrated into the process, optimising both the ecological and economic impact of a project. The certification does not assess individual measures separately, unrelated from each other, but focuses on their holistic performance. Nonetheless, a minimum performance must be achieved in all relevant sections to obtain a certificate.

This approach allows the DGNB system to be more adaptable to local conditions of a project, as it does not impose a fixed catalogue of measures onto a project, like other certification systems do, but allows the planning team to find a unique solution within the given context of a project.

The life cycle of a project is accompanied by an integral planning team from project development until the project is demolished and recycled. All planning disciplines are working together to shape the development together. Urban design, architecture, mobility, landscaping and many other disciplines are interwoven through complex interdependencies. Integral planning makes these interdependencies transparent to all stakeholders and allows for simultaneous and iterative optimisation.

The minimum size of an urban district for DGNB certification is 2 ha of gross development area. The district needs to consist of more than two buildings and at least two plots to be developed. It is also essential that there are public or publicly accessible spaces and infrastructure on site. In terms of usage, urban districts need to be of mixed use, requiring a residential element of at least 10% and at most 90% of the gross floor area (DGNB, 2012). Last, but not least, all areas within a certification district need to be assessed and the boundary of the district must be confirmed before the certification process starts (see Fig. 19.1).

Fig. 19.1 Example of the boundaries of an urban district for DGNB certification (*Source* DGNB, 2012)

19.3 Evaluation

The evaluation of urban districts focuses on the publicly accessible spaces between the buildings. This includes traffic routes, squares, and green spaces. In addition, the general solutions for managing energy, water, and waste are being considered in the evaluation.

Another important factor is the regulatory framework for the development of buildings in the district. Whereas individual buildings within the development do not need to be DGNB certified, their base values are considered during the certification process in terms of e.g. energy-, water- and heat-usage.

Furthermore, evaluation is not limited to the site itself. Its surroundings are also part of the assessment: Neighbouring infrastructure (e.g. a train station), services (e.g. hospitals, schools) and natural assets (e.g. parks) can all be considered (see Fig. 19.1).

The urban district certification uses the same sections as DGNB single building certification to assess the quality of a project: environmental quality, economic quality, socio-cultural and functional quality, technical quality and process quality. Contrary to the single building certification, in the urban district certification, site quality has influence on all the other quality sections (compare Fig. 19.2).

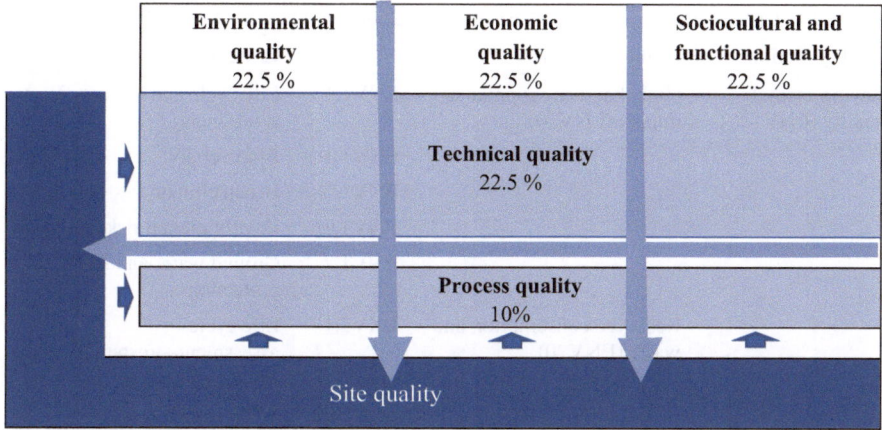

Fig. 19.2 Six quality sections define DGNB certification for Urban Districts, contrary to the single building certification, site quality has an influence on all the other criteria (*Source* DGNB, 2012)

These six quality sections are split up in eleven subgroups, each with a number of criteria, assessing e.g., the amount and quality of public spaces and 'placemaking', i.e. the definition of public spaces with a specific identity, in the area, the area's contribution to the municipal economy, the involvement of local actors in the development plan, the social and functional mix in the area and many others (see Table 19.1). Each of the parameters has a different weight, leading to a total score (as a percentage) that defines the sustainability score.

19.4 Certification Process

Urban districts often develop in several phases over a long period of time and the construction phase is considerably longer than is the case for individual buildings. The process leading to DGNB certification reacts to this by using three instead of two certification stages (see Fig. 19.3).

The pre-certificate is followed by two separate certificates. The first of these, with a term of five years, can be awarded when 25% of the infrastructure has been completed. The second serves as the final certificate, as it is unlimited in duration. It is awarded when 75% of the buildings have been completed.

Table 19.1 Criteria used to rate DGNB-certified developments (*Source* DGNB, 2012)

Quality sections	Evaluation topics	Nr.	Criteria
Environmental quality (ENV)	Global and environmental impact (ENV10)	ENV1.1	Life cycle and impact assessment
		ENV1.4	Biodiversity
		ENV1.5	Urban climate
		ENV1.6	Environmental risks
		ENV1.7	Ground water and soil protection
	Resource consumption and waste (ENV20)	ENV2.1	Life cycle assessment—resource consumption
		ENV2.2	Water cycle
		ENV2.3	Land use
Economic quality (ECO)	Life cycle cost (ECO10)	ECO1.1	Life cycle costs
		ECO1.2	Local economic impact
	Creating value (ECO20)	ECO2.1	Resilience and adaptability
		ECO2.3	Land use efficiency
		ECO2.4	Value stability
Sociocultural and functional quality (SOC)	Health, comfort and user-friendliness (SOC10)	SOC1.1	Thermal comfort in open spaces
		SOC1.6	Open space
		SOC1.9	Noise, exhaust and light emissions
	Functional quality (SOC20)	SOC2.1	Barrier-free design
	Social quality (SOC30)	SOC3.1	Urban design
		SOC3.2	Social and functional mix
		SOC3.3	Social and commercial infrastructure
Technical quality (TEC)	Infrastructure (TEC20)	TEC2.1	Energy infrastructure
		TEC2.2	Resource management
		TEC2.4	Smart infrastructure
	Transport (TEC30)	TEC3.1	Mobility infrastructure—motorized transportation
		TEC3.2	Mobility infrastructure—pedestrians and cyclists
Process quality (PRO)	Process quality (PRO10)	PRO1.2	Integrated design
		PRO1.7	Consultation
		PRO1.8	Project management

(continued)

Table 19.1 (continued)

Quality sections	Evaluation topics	Nr.	Criteria
		PRO1.9	Governance
	Quality assurance during utilization phase (PRO30)	PRO3.5	Monitoring

Fig. 19.3 The process of DGNB-district-certification is split into three steps (*Source* DGNB, 2012)

19.5 Inclusive Quarters with DGNB

The DGNB system does not require specific technical solutions to problems the elderly face exclusively, but rather solutions that serve all the people using a and living in a city district. This, however, means that there are criteria that cannot be met without considering the needs of the elderly.

It is also important to understand that there is no prototypical "old person" whose needs are to be met, but that it is necessary to develop open-ended, adaptable solutions for all types of users. In the following paragraphs, a couple of criteria and their effects on inclusion will be presented as examples.

19.5.1 Social- and Functional Mix (SOC 3.2, SOC 3.3)

The direct environment surrounding the home is especially important for the elderly, as they often have to cope with reduced personal mobility.

DGNB requires a social and commercial infrastructure within the urban district, which includes certain services (e.g. doctors, hairdresser) and facilities (e.g. stores, restaurants, church) in close proximity to one's residence. This enables an elderly person to fulfil their daily life within the familiar framework of their own neighbourhood.

A wide array of different types of residences should be offered within a district to allow a mix of different types of households and of different age and social groups.

This gives old people the chance to stay within their urban district, even if their requirements in terms of housing change.

If a person is no longer able to live independently within one's own residence because of poor health, especially dementia, the availability of small-scale and decentralised support-oriented housing forms to which they can move allows people to keep some degree of independence. If any are available close to their original residence, social connections to friends and family can be kept intact. Furthermore, orientation within a well-known environment is easier. At the same time, diverse options in terms of housing also make it easier for family members to find housing opportunities nearby if necessary.

In cases where families are not living close or families have no children, special institutions, like e.g. the Multi-Generation-Houses, can serve as a meeting point between different age-groups and also social groups, and allow people to form social connections within their community.

DGNB certification also awards extra points for the inclusion of a sociologist into the integral planning team (PRO 1.2).

19.5.2   Public Spaces (SOC 1.6, SOC 3.1)

As important as the personal residence is for the elderly, 'living at old age' should not be reduced to 'residing at old age'. High quality and diverse public spaces play a very important role in activating the life of all people living in a neighbourhood.

As personal mobility is reduced with age, areas between one's residence and the places of daily life should offer opportunities for rest. At the same time, generous landscaped areas can invite walks or even offer gymnastic areas, thereby enhancing health through physical activity. As the public life in the quarter is organised around urban plazas, parks and playgrounds, possibilities for participation in the public life are offered to all residents.

The layout of the district should also be considered, as big streets with a lot of traffic can act as barriers and hinder movement. This is especially true for the elderly that may move more slowly and have reduced sensory capabilities to cope with a busy street.

DGNB certification requires a landscape architect as part of the integral planning team, but also encourages the integration of e.g. artists to further enhance the public areas (PRO 1.2).

19.5.3   Accessibility (SOC 2.1)

Accessibility is the basis for the continued ability to satisfy one's basic needs. It is important to not only consider accessibility in terms of a bodily handicap, but also include other types of mental or sensory handicaps. As dementia and Alzheimer's increase with rising life expectancy, the specific limitations of these illnesses have to be accounted for.

The German Federal Minister for the Environment, Nature Conservation and Nuclear Safety has developed a set of guidelines for accessibility ("Leitfaden Barrierefreies Bauen"), which can serve as a basis for the integration of best practices in urban design and architecture (BMU, 2017).

Accessibility should be deeply integrated in the basic circulation concept and allow everybody to use the same pathways throughout the district, as this leads to an added sense of security. Signage systems should support the accessibility concept and be designed keeping in mind people with sensory or cognitive limitations.

19.5.4 Thermal Comfort (SOC 1.1)

Heat waves can have significant impact on individual health and present a challenge for public health services. Due to changes occurring in the thermoregulatory system, the elderly are particularly vulnerable to such episodes. It has been shown in epidemiological studies that heat-related mortality rises with increasing age from around 50 years onwards (Kovats & Hajat, 2008: 46–47).

Further, one of the more certain impacts of future anthropogenic climate change will be an increase in heat waves, and these will become more intense. Research has found that the mortality rate during a heat wave increases exponentially with the maximum temperature, an effect that is exacerbated by urban heat islands.

These urban heat islands are areas of a city that are significantly warmer than their surroundings due to human construction and activities. The main cause of the urban heat island effect is the modification of land surfaces and a lack of vegetation. Waste heat generated by energy usage is a secondary contributor.

Mitigation strategies include providing green roofs, planting trees, reducing hard surface areas, as well as integrating water features, outdoor shading or other devices to actively reduce surface temperatures. Orientation and layout of the district is of additional importance, as solar shading by buildings and air streams between them have to be considered. DGNB certification awards additional points for the integration of an urban climate specialist that can support the integral planning team (PRO 1.2).

19.6 Outlook: Adaptation to China

The DGNB system can be adapted internationally. As it is goal and performance oriented and not based on a specific legal context, the system for single building certification has been adapted to the local environment of countries such as Austria, Switzerland, Denmark, Bulgaria and China. A number of DGNB-certified single building designs have already been completed in China, e.g. the German Enterprise Centre in Qingdao.

Since its introduction to the market in 2012, the neighbourhood certification programme has been used successfully both in Germany and abroad, now being the most used Neighbourhood Sustainability Assessment tool in Europe. Contrary to the single building certification, the DGNB urban district certification system has not yet been implemented in China. As the requirements of documentation for district certification are less reliant on specific product data than the single building certification, the adaptation to other countries is generally less complex.

China is facing two major societal trends right now: the migration of the rural population towards urban centres and its ageing society. As stated in the introductory chapter of this text, DGNB certification is a good tool to tackle both topics.

As the size of urban agglomerations rises, the need for sustainable development becomes more and more prominent. Looking at the urban centres in China, energy-use, mobility, and pollution are already sizeable problems at this moment. Neighbourhood Sustainability Assessment tools, like DGNB, can help to achieve sustainability goals while extending the urban fabric.

An ageing society also introduces specific challenges to urban planning in China. An in-depth study by OECD (2014) that cited policy challenges in Beijing shows the need to plan urban spaces specifically with old people in mind. Building new social housing, as well as providing stable communities with facilities that cater to the needs of the elderly will be essential. As we have seen, DGNB certification, as a tool, addresses these questions and allows for evaluating different options during the design phases.

In conclusion, DGNB district certification can be a useful tool in generating sustainable, inclusive urban districts. Its focus on being adaptable and open-ended helps to find specific local solutions and offers an opportunity for its application in China.

References

BMFSFJ, Bundesministerium für Familie, Senioren, Frauen und Jugend. (Ed.) (2000). Dritter Altenbericht. Alter und Gesellschaft. Stellungnahme der Bundesregierung. Bericht der Sachverständigenkommission. Retrieved from: https://www.bmfsfj.de/bmfsfj/service/publikationen/3-altenbericht--95592 [04.05.2021].

BMU, Bundesministerium für Umwelt, Naturschutz, Bau und Reaktorsicherheit (Ed.) (2017). Leitfaden Barrierefreies Bauen. Hinweise zum inklusiven Planen von Baumaßnahmen des Bundes, Berlin. Retrieved from: https://www.bbsr.bund.de/BBSR/DE/veroeffentlichungen/ministerien/bmub/verschiedene-themen/2017/leitfaden-barrierefreies-bauen.html [04.05.2021].

Brech, J. (2004). Wir werden immer älter - Sind die richtigen Fragen schon gestellt? In BauWohnberatung Karlsruhe; Schader-Stiftung (Ed.), *Neues Wohnen fürs Alter. Was geht und wie es geht* (pp. 38–46). Anabas.

DGNB, Deutsche Gesellschaft für Nachhaltiges Bauen e. V. (Ed.) (2012). Neubau Stadtquartiere, DGNB Handbuch für nachhaltiges Bauen (Version 2012) (p 62). Stuttgart.

Heye, C., & Wezemael, J. E. (2007). Herausforderungen des sozio-demographischen Wandels für die Wohnbauindustrie. *DisP—The Planning Review, 43*(169), 41–55. https://doi.org/10.1080/02513625.2007.10556981

Kovats, R. S., & Hajat, S. (2008). Heat stress and public health: A critical review. *Annual Review of Public Health, 29*(1), 46–55. https://doi.org/10.1146/annurev.publhealth.29.020907.090843

Mollenkopf, H., Oswald, F., Wahl, H.-W., & Zimber, A. (2004). Räumlich-soziale Umwelten älterer Menschen: Die ökogerontologische Perspektive. In A. Kruse & M. Martin (Eds.), *Enzyklopädie der Gerontologie* (pp. 343–361). Verlag Hans Huber.

OECD, Organisation for Economic Cooperation and Development. (2014). Local scenarios of demographic change—The silver and white economy: The Chinese demographic challenge, Paris (pp. 22–34). Retrieved from: https://www.oecd.org/employment/leed/oecd-china-report-final.pdf [21.10.2020].

OECD, Organisation for Economic Cooperation and Development. (2015). *Ageing in cities*. Paris (p. 5). Retrieved from: https://www.oecd.org/regional/ageing-in-cities-9789264231160-en.htm [05.06.2019].

UN DESA, United Nations Department of Economic and Social Affairs, Population Division. (2016). *The world's cities in 2016—Data booklet* (p. 2). https://doi.org/10.18356/8519891f-en

UN IEA, United Nations International Energy Agency. (2017). *Towards a zero-emission, efficient, and resilient buildings and construction sector*. Global Status Report 2017 (p. 14). Retrieved from: https://www.worldgbc.org/sites/default/files/UNEP%20188_GABC_en%20%28web%29.pdf [30.01.2020].

World Bank. (2010). *Cities and climate change: An urgent agenda* (p. 23). Retrieved from: https://openknowledge.worldbank.org/bitstream/handle/10986/17381/637040WP0Citie00Box0361524B0PUBLIC0.pdf?sequence=1&isAllowed=y [04.05.2021].

Christian Eichinger has been working at KSP Engel Architekten since 2009. He is currently holding the role of the Munich Office head and is an authorised signatory. From 2013 to 2017, he served as the International Director of Design for projects in the Beijing office. He studied Architecture at the Technical University of Munich and the Illinois Institute of Technology, Chicago, USA.

Chapter 20
Digital Emotion-Mapping for Barrier-Free Design. How User-Specific Assessment of Environmental Experience Helps to Design Better Cities

Sebastian Schulz and Zheng Chen

20.1 Introduction

20.1.1 Perception of Urban Environments

Predicted human experiences and environmental psychology are key factors in designing and evaluating urban environments today. Additionally, recent studies revealed that better scenic views in public spaces correlate with lower discomfort and lower stress levels of citizens (Capaldi et al., 2014; Florian et al., 2011; Haluza et al., 2014; Lafortezza et al., 2009; Maas et al., 2006; Seresinhe et al., 2015 and Velarde et al., 2007). This suggests visually comfortable and easy-to-orientate urban spaces may have significant impact on peoples' well-being. Therefore, environmental psychology matters, not only for functionality and comfort in the design of spaces, but also for immediate health of urban citizens. This becomes even more important in rapidly ageing societies with increasing health care demands, which we can find in Germany and (urban) China.

In this context, recording and measuring in-situ human experiences in actual environments is a new field to explore. Emotions and feelings always accompany humans. There are multiple reasons for a person's emotional state (and impacts on health), which might not always be linked to one's visual experience per se. Different

Z. Chen
Tongji University, CAUP, Yangpu, China

S. Schulz (✉)
Transport Planner, General Manager, Buero stadtVerkehr Planungsgesellschaft mbH & Co. KG, Hilden, Deutschland
e-mail: schulz@buero-stadtverkehr.de

© The Author(s) 2025
Deutsche Gesellschaft für Internationale Zusammenarbeit (GIZ) GmbH et al., (eds.),
Sustainable Aging, https://doi.org/10.1007/978-3-662-69139-7_20

user groups perceive and rate urban space differently based on their personal needs and objectives.

Designing spaces that appeal to all is therefore a most challenging task. However, designing inclusive spaces, which can be properly used by all age groups, from children to the elderly, is a key concern in contemporary urban planning. Guidelines for accessible design or completely integrated barrier-free systems have been defined, and are now playing a key part in sustainable evaluations of buildings and cities (i.e. in sustainable certification systems, such as DGNB, LEED, Chinese GBL). Yet, engaging the public or targeted user groups in design processes, often does not produce the best outcomes. Public participation experiences in Germany have shown that opinions and decisions for planning, designing or evaluating urban spaces are often driven by strong emotions of particular interest groups, rather than objective and equally weighted facts.

How to balance the spatial needs for everyone in the urban context? How do users perceive urban environments, and which spatial features contribute most to the well-being of citizens? How can we best design inclusive, barrier-free cities, which are convenient, attractive—and most of all: accepted—by all users? These are the underlying questions, urban planners in Germany, China, and world-wide, have to find objective answers to.

20.1.2   Competition Over Public Space

The arrangement and design of public space has always been a compromise of different types of use and occupations. Car-dependent development over the past decades has transformed the spaces between buildings not only in size, but also on its basic functions; from spaces of interaction and trade to somewhat simplified transport corridors.

Functional zoning of city space has largely encouraged this transformation. Today's modern streetscapes often feature wide roads of multiple lanes, pure transport corridors with physical barriers obstructing pedestrians or cyclists to interfere with car traffic, allowing free passage wherever possible to increasing motorised traffic. The movement in public space in today's cities can sometimes evolve into adventurous detours. Different user groups of car-riders, bike-riders, and pedestrians unavoidably interfere at crossings and junctions, making these places the most dangerous spots for all users competing for their rightful moments of passage. Especially for mobility-impaired and elderly people, these challenges become serious obstructions from public life (see Figs. 20.1 and 20.2).

In terms of human emotions, it has been proven that stress levels and comfort/discomfort perception highly correlate with traffic conditions and design of public space (Bergner et al., 2013). Several studies in recent years have therefore discussed the proposals of re-implementation of traditional elements and design aesthetics in Chinese urban and landscape planning for the purpose of creating healthy and liveable places (Chen & Thwaites, 2013; Hassenpflug, 2013; Wang & Meng, 2015; Wang &

Fig. 20.1 Competition over public space in Shanghai (*Source* Schulz)

Fig. 20.2 Shared-bikes in Shanghai. Convenience for whom? (*Source* Schulz)

Ruan, 2015). In the process of smart city approaches in China, the understanding of people's well-being and interaction with their urban environments have gained unprecedented importance in contemporary planning practice.

Also, in Germany, the topic of people-friendly urban spaces has become a major concern for city administrations, as a rapidly ageing population has different require-ments towards urban mobility and comfort. Several regulations and guidelines for barrier-free design have been issued to develop cities and towns towards integrated and comfortable places for all user groups. But as examples in Germany sometimes reveal, simply following guidelines and laws can still lead to misunderstandings, confusion and consequently discomfort for affected user groups (see Fig. 20.3).

For designing urban landscapes that are truly comfortable for all user groups, including the elderly and mobility-impaired, more research on environmental psychology and in-situ assessment of emotions is necessary.

Fig. 20.3 Bus stop in Germany with applied barrier-free guidelines, yet the targeted user groups avoid the place because of its steep slope (*Source* Schulz)

20.1.3 Quantification of Human Emotions

Studies on the empirical value of emotion mapping has highlighted possible symbioses of individual perception of spaces on the one side and urban design on the other side (Kwan 2007). There are currently three developed methods to document environmental experiences. The most used method is the conduct of phenomenological interviews (Creswell, 1998; Deming & Swaffield, 2011; Groat & Wang, 2002), in which participants are asked to recall details of certain, more subjective, incidents as adequately as possible.

An alternative assessment method is psychological sales, based on subjective user feedback supported by statistical analysis. Another, rather new method with the increasing growth of digital sensor technologies (Hartig et al., 1997; Tzoulas et al., 2007), is to record biological conditions using wearable biosensor devices (Aspinall et al., 2013; Bergner et al., 2013; Chen et al., 2016; Roe et al., 2013 and Wang et al., 2016).

In our initial experiment, all above methods have been applied, with a particular focus on biosensor technology. The study is especially designed to check and demonstrate the feasibility of advanced biosensor application in urban environments in comparison with other methodologies, as described above. The ultimate long-term goal of this research is to build a design decision support system, which includes quantified information from viewpoints of all potential user groups.

20.2   Methodology

20.2.1   Emotion Mapping and Bio-Sensory Data

Emotions are complex to understand and therefore difficult to measure, but recent advancements in cognitive neuroscience offered new opportunities. Emotions can be understood in several dimensions, i.e. "valence" (biphasic emotion, whether and how much one likes or dislikes something), "arousal" (a neither positive nor negative indicator for neurological stimulation) and arguably a third approaching/avoiding motivation dimension (Bradley & Lang, 2006; Mauss & Robinson, 2009). Most bio-sensory measures fall into "valence" and "arousal" dimensions (Lang, 1995). The state of arousal ranges from neutral to highly aroused or emotionally triggered. The valence level as a biphasic placement of either pleasant or unpleasant reaction, completes a measurable set of indicators for either positive or negative emotional response. For measuring arousal and valence levels, several indicators, such as skin conductance, skin temperature, electroencephalogram (ECG), and electromyography of facial muscle expression (EMG) are being assessed according to their measured value change rates (see Fig. 20.4).

When change rates of biosensors correlate with each other (compare Bergner et al., 2011, 2013; Bradley & Lang, 2006; Lang, 1995), they can indicate emotional reactions towards a specific view or situation. This offers an opportunity to specify emotions in both valence and arousal, which therefore differentiates positive emotions (excitement) from negative ones (stress/fear). To urban planners and landscape designers, this information can be crucial to evaluate and determine healthy and sustainable places.

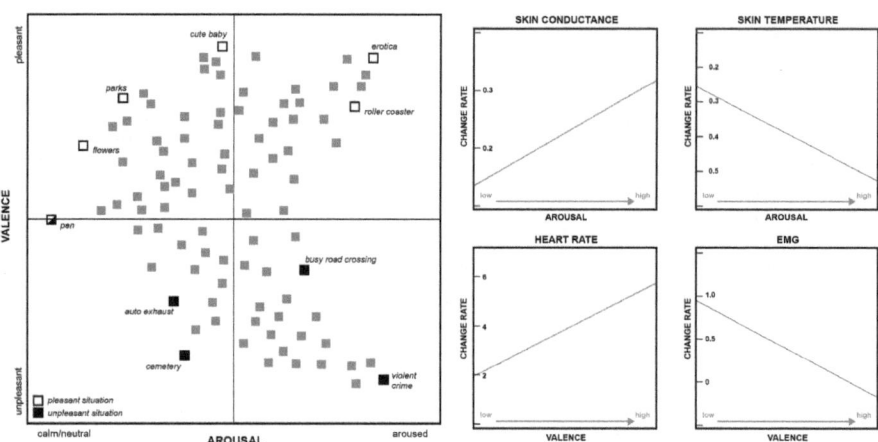

Fig. 20.4 Arousal and valence indicator assessment (*Source* own graphic based on Lang, 1995) and expected correlation with bio-measurements

20.2.2 Pilot Study in China

The pilot study was designed to test the feasibility of measuring emotions, affective valence and arousal levels to be specific, using a portable multi-channel physiological device during an in-situ walk in the real environment. Supported by empirical evidence from literature, we expected to measure affective valence using heart rate and facial muscle EMG, while to measure affective arousal using skin conductance, skin temperature and ECG based on the findings of Lang et al., 1993; Lang, 1995; Mauss & Robinson, 2009.

For the experiment's location, a 15-min walking route across the Tongji University Campus in Shanghai was chosen. The selection of the university campus was on clear purpose, as the route was expected to vary for environmental experiences in a walkable range. The proposed route across the campus can be divided into eight zones or locations. It consists of open spaces as well as dense alleys, green spaces and concrete plazas, busy roads with traffic and rather calm pedestrian walkways (see Fig. 20.5).

Several participants at the university (students and teachers) between the age of 20–40 years were recruited by the project leaders and asked to walk the selected route in the same direction for three times, with small breaks between each time. They were asked to perform the walk in a stable and modest walking speed, so that exhaustion and impacts on measured data were kept on a minimum impact level. At each walk, they were carrying a 6-channel Procomp Infiniti device, a video-camera attached to the head gear, as well as a GPS device. The whole process of the walks was also videotaped by a researcher walking behind the participants.

Description

1 Small street with trees in between sporting grounds

2 Small footpath through green space with monument

3 Footpath between tall buildings and parking lots

4 Major road crossing, medium traffic

5 Sino-German campus plaza with wide open spaces between buildings

6 Dormitory buildings alley

7 Street through green space area

8 Small alley alongside dormitories and green spaces

Fig. 20.5 Chosen route with eight different perception areas (*Source* Schulz and Chen)

All participants were familiar with the location. Therefore, stress reactions determined to orientation, navigation, or "unexpected visual surprises", could be reduced to a minimum. Interviews after the experiment confirmed that no participant had any difficulties with the route itself or for general orientation during the walk.

20.2.3   Application

As reactions and perceptions cannot be fully regarded to visual experience alone, other influences, such as noise, smell, weather conditions, also play a role. Except the latter, however, all aspects that might influence human perception, are also important to be considered in urban planning and design. To rule out special or individual circumstances leading to positive or negative reactions during the experiment, each participant was asked to state his personal conditions (mood, health, etc.) in advance, and write a brief narration for each walk afterwards, specifying visual experiences—whether negative or positive—and other happenings or interactions.

In this context, phenomenological interviews are the most used method for environmental psychology experiments (Beidler, 2007; Callahan, 2000). In this study, the participants were asked to use six kinds of labels to mark properties at the different zones and their emotions. Participants were then being interviewed for more details (what had happened, how they felt, etc.) about those marked zones (see Figs. 20.6, 20.8).

Additionally, for psychological scales analysis, participants were asked to grade marked spots, which were selected based on the route design and general impression. The spots were assessed in arousal and valence. Each of the two dimensions were scored from grade 1 to 9. The product of average arousal and valence of each spot was used to assess the level of emotion (the higher the score, the more positive the emotion, see Fig. 20.7).

20.2.4   Biosensor Mapping

Data collection with digital bio-mapping devices underlie certain preconditions. Usually, multisensory data experiments take place under enclosed (indoor) lab conditions (Tost et al., 2015), where environmental impacts or external effects are minimal on the measured results. For outdoor environments, several aspects have to be pointed out, which interfere with human wellbeing and comfort: weather, wind, temperature, air humidity, light/shadow, smell or noise, among others. Distortions by such external influences can only be ruled out with a certainly high number of datasets under different conditions of the outdoor environment, or by other control mechanisms, such as interviews or statistical evaluations (as described above).

Another, more difficult obstacle for data collection is the activity on an outdoor walk itself. An ongoing walk impacts heart rate/pulse and other biological indicators

1. The spots where you strongly feel pleasant, relaxing and/or comfortable.
2. The stretches that you find appealing.
3. The directions where you enjoy a view.
4. The way-finders that might be attractive or unattractive.
5. The places or stretches where you are unpleasant or anxious.
6. The sponts where you are particularly unpleasant and/or anxious.

Fig. 20.6 Affective map via phenomenological Interview (*Source* Schulz and Chen)

used in this study. For example, increases in ECG measurements and stress reactions towards the end of the route, could be very well linked to the activity or even to the proband's knowledge (and arousal) that the "finish line" of the walk is coming close. Some test runs also indicated that the walking speed increased towards the end of the route for almost all probands, which also depicted effects on out-of-the-ordinary arousal measurements (see Fig. 20.7).

20.2.5 Data Processing and Evaluation

Post-processing of the recorded datasets needs to be undertaken prior to emotion analysis. In a first step, collected data indicators were normalised from their individual sample rate (i.e. 256 samples per second) to a comparable level of 1 s each. This enables biosensor data to be synchronised with the 1 s interval of GPS tracks. The

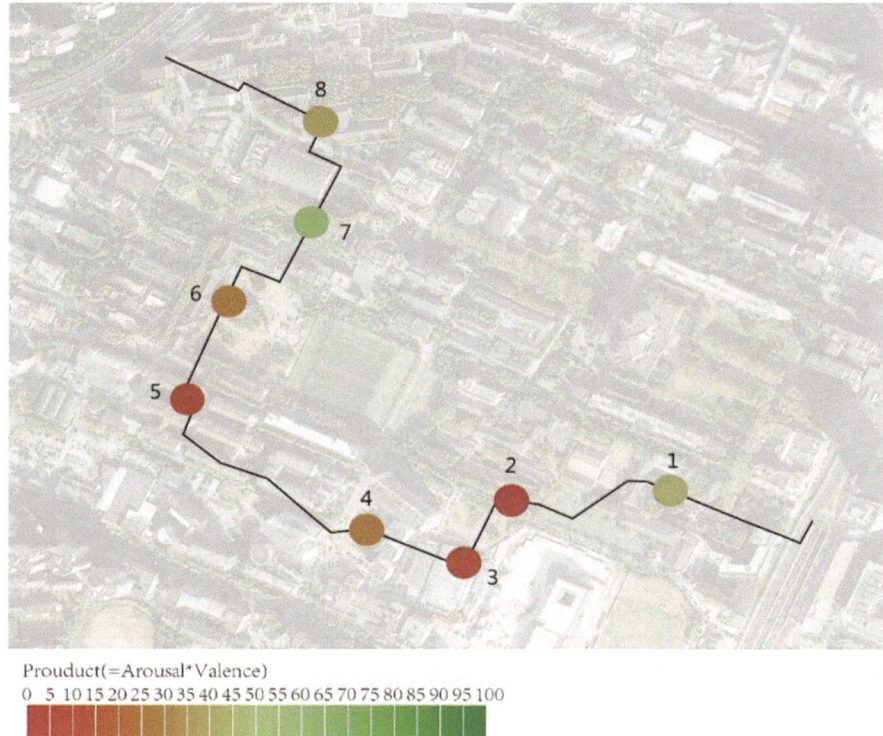

Prouduct(=Arousal*Valence)
0   5 10 15 20 25 30 35 40 45 50 55 60 65 70 75 80 85 90 95 100

Fig. 20.7 Affective map via psychological scales (*Source* Schulz and Chen)

average rates are still consistent and depict data in appropriate precision for the project purpose. However, not all biosensors are possible to be averaged at this stage properly. ECG, EMG and electroencephalogram (EEG) data were recorded on several different frequency channels, which need complex resampling and filtering via MatLab software in the post-processing stage.

For locating arousal and valence in data channels on GPS-tracked routes, the focus is put on change rates of single indicators. The correlated increases or decreases for arousal and valence indicators show the emotional responses of either pleasant or unpleasant feelings of the participants. After thorough analysis of the usable datasets of each walk and each participant, they were depicted via the time- and GPS-stamp in a GIS-based heat map (see Fig. 20.8). According to this map, seven meaningful hot spots for emotional responses can be identified alongside the route, three of which with unpleasant and three with mostly pleasant emotional responses. A seventh hot spot depicts a negative response followed by stronger positive responses. This can be explained by the change of environments in this area, as participants walked first in between parked cars and two buildings, then turned to continue to walk adjacent to a park.

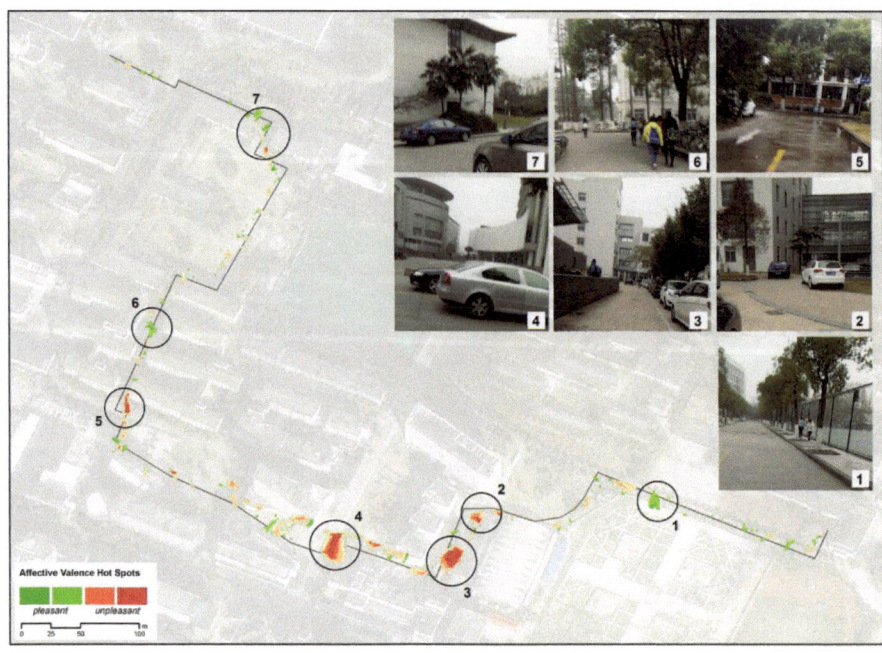

Fig. 20.8 Affective valence heat map (*Source* Schulz and Chen)

For all identified hot spots, unpleasant emotions show comparably higher magnitudes than pleasant ones. Along the route, further minor triggers can be found, both for positive and negative emotions, which after viewing video records and interviews can be linked to occasional interactions or in some cases be the result of visual perception, as interviews showed. Positive triggers often appear in areas, where a long-distance line of sight and good general overview of the surroundings is given, while minor negative triggers appear in dense surroundings. Confirming the initial hypothesis of this study, unpleasant emotional responses are also typically located at spots of high traffic or traffic-related interactions (road crossings, parking lots), while positive emotions or pleasant reactions can be found in rather calm and spacious areas along the route.

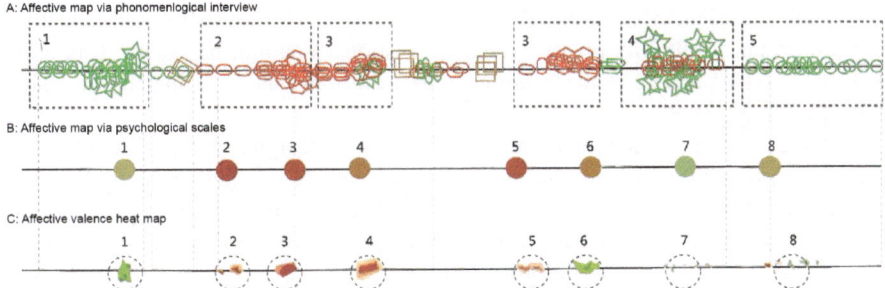

Fig. 20.9 Comparison among the results of all three methods (*Source* Schulz and Chen)

20.3 Results

20.3.1 Comparison of Applied Methods

The comparison of bio-sensory measurement and the two rather traditional methods indicated that the newly developed method of emotion mapping is valid. It can be found that there are strong similarities in determining hot spots in all three methods (see Fig. 20.9).

Comparing to stated narratives and phenomenological interviews, certain spots surprisingly did not match with results from biosensor indicators. Almost all participants named zones 2, 3 and 5 as the most unpleasant. On the opposite side, zones 1, 7 and 8 were named as the most pleasant, both for walkability and visual aesthetics. The measured results depicted in the heat map though, do not confirm strong feelings or emotional responses at all allegedly unpleasant or pleasant locations. Zone 5 on the one hand, was stated as the most negative perception by all participants, naming specifically the rather dark walkway in between tall and grey buildings. In data rows and hot spot analysis, slightly unpleasant emotions can be verified, but not to an extent, which would be comparable to other unpleasant hot spots linked to interactions and traffic (compare zone 4).

Similar to these findings, zone 7 was judged as a spot with most positive visual aesthetics by the participants. However, it does not show significant impacts on pleasant emotional responses in biosensor measurements. Minor magnitudes exist in the hot spot analysis, but do not correlate with the strongly identified hot spots from stated narratives and interviews.

20.3.2 Interpretation and Implications for Design

The results of this initial case study indicate that emotional responses towards urban environments are more likely impacted by interactions to a much larger magnitude than visual experiences of a place. Therefore, is seems that direct linear conclusions between urban design and positive/negative responses cannot be drawn. It is more likely that consequences from design need to be further analysed, instead of design aesthetics themselves.

Traffic and interaction density generated by space design are important topics to consider in planning processes and should not be neglected in favour of architectural design topics. Direct access (no detours), pedestrian-oriented, barrier-free surfaces and easy orientation need to be further elaborated by urban planners, especially in context of inclusive design for mobility-impaired people. Places of anxiety, dark areas and confusing spaces are to be avoided, traffic nodes, street crossings and walkways need to primarily consider the needs of the weakest instead of convenience for the majority.

For practical application, this could mean for urban planners and decision-makers to pay more attention to perception impacts for users when planning spaces. As a result of the pilot study, several small-scale alterations have been proposed to increase walkability and well-being, while still complying with standards and regulations; i.e. the arrangement of parking spaces on small alleys can be re-organised to benefit all users by increasing safety and comfort (see Fig. 20.10). In general, whether designing new districts or retrofit/improve existing neighbourhoods, results from emotional studies have the potential to largely support the design and decision-making process.

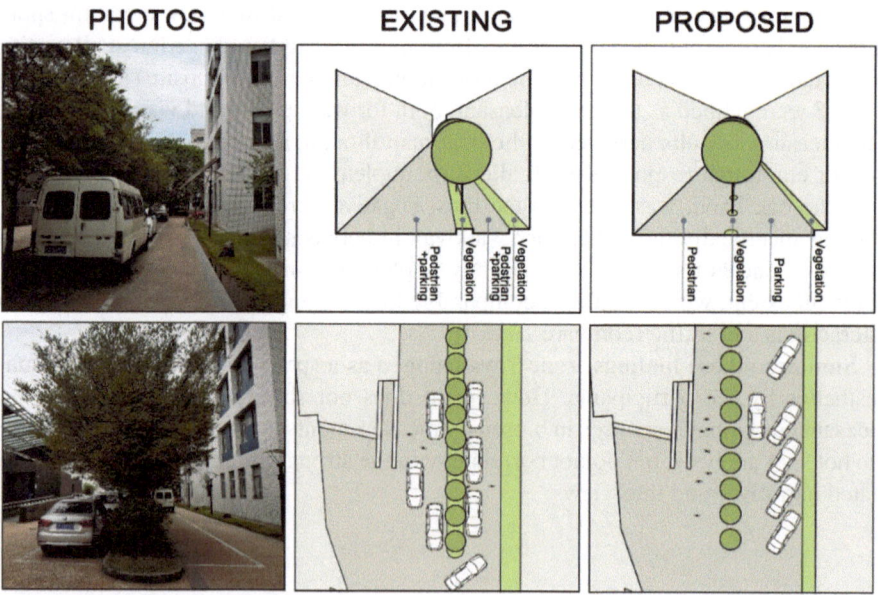

Fig. 20.10 Small-scale design changes for improved well-being (*Source* Schulz and Chen)

The method of bio-sensory measurements shows its potential on this subject, especially when compared to the other methodologies. Firstly, with wearable bio-sensory devices, in-situ details can be captured better, and probably to more truth-revealing emotions rather than through post-experience, subjective recollections. What's more, is a holistic tracking of biological emotions, which at the time of their immediate recording are not consciously or unconsciously interpreted by the proband. Through bio-sensory measures, perception of places can be assessed before opinions are formulated and brought to discussion in design participations. This approach of design decision-making could benefit the often-unheard voices of the weakest user groups.

20.4   Conclusion

The case study showed first possibilities multisensory data analysis can have on the characters and designs of urban spaces. However, many difficulties and challenges exist for proper data interpretation and application. First of all, the tested device produces data of varying quality under outdoor conditions and activities, which has to be reviewed in rather time-consuming post-processing procedures. The devices used in this pilot study, are also not suitable for common every-day data collections, therefore is this study not fully replicable to practical design decision-making. Many obstacles remain for proper data collection and analysis.

Nevertheless, with the appropriate steps taken, multisensory data is able to not only identify stressors (as wristbands do) or—in case of urban space—locations with unpleasant emotional reactions but can also depict possible positive connotations towards the built environment. As presented, positive and negative triggers can be identified, even though they could not undoubtfully be linked to visual perception and design experience. There are several additional factors and influences to consider in open urban environments, which might have a higher impact on emotional responses and bio-sensory data rather than visual aesthetics.

Age group and user group specific analysis is a further step for the project, since different user groups perceive their surroundings differently. The pilot run has shown no significant difference in perception between age groups. However, more samples from different social backgrounds and especially elderly or mobility-impaired probands are of further interest. In addition, cultural differences in the perception of space are another issue to be tackled. Do elderly and impaired people in Germany or China have different preferences and perceptions? For truly designing urban spaces including the necessities and needs from potentially all user groups, more experiments have to be conducted and more datasets have to be evaluated.

Since technological development and ongoing enhancement of multi-biosensor devices persist, the findings of emotional responses towards environments and sceneries will gain importance and hold their share for designing more liveable urban landscapes in both Germany and China. A design supporting system can be built with the help of affective mapping, clearly depicting environmental stressors as well as potential attractors. According to this, designers can improve urban spaces that might contribute to negative emotions. Besides, following examples of positive response environments, designers can consolidate the sense of place in their work. Modern computing programmes, which evaluate spaces based on pre-specified algorithms are already in development.

On a long-term perspective, large scale sample collection on human emotional responses in different places and artificial intelligence systems can be combined to complex, yet more efficient (and fair?) evaluations for sustainable, integrated urban design in comparison to currently available methodologies of participation and decision-making processes.

Acknowledgements This project is funded by the National Natural Science Foundation of China (51408429), Shanghai Pujiang programme (14PJC099) and the Fundamental Research Funds for the Central Universities, Key Laboratory of Ecology and Energy-saving Study of Dense Habitat (Tongji University), Ministry of Education, and Tongji Architectural Design (Group) Co., Ltd.

References

Aspinall, P., Mavros, P., Coyne, R., & Roe, J. (2013). The urban brain: Analysing outdoor physical activity with mobile EEG. *British Journal of Sports Medicine 0,* 1–6.

Beidler, K. (2007). *Sense of place and New Urbanism: Towards a holistic understanding of place and form.* Ph.D. 3310170, Virginia Polytechnic Institute and State University, United States—Virginia. Retrieved from: http://proquest.umi.com/pqdweb?did=1674955691&Fmt=7&clientId=36305&RQT=309&VName=PQD, December 12, 2020.

Bergner, B., Zeile, P., Papastefanou, G., Rech, W., & Streich, B. (2011). Emotional barrier-GIS—A new approach to integrate barrier-free planning in urban planning processes. In M. Schrenk, V. Popovich, & P. Zeile (Eds.), *Proceedings REAL CORP 2011* (pp. 247–257). Essen.

Bergner, B., Exner, J.-P., Memmel, M., Raslan, R., Taha, D., Talal, M., & Zeile, P. (2013). Human sensory assessment methods in urban planning—A case study in Alexandria. In M. Schrenk, V. Popovich, P. Zeile, & P. Elisei (Eds.), *Proceedings REAL CORP 2013* (pp. 407–417). Rome.

Bradley, M. M., & Lang, P. J. (2006). Motivation and emotion. In J. T. Cacioppo, L. G. Tassinary, & G. Berntson (Eds.), *Handbook of Psychophysiology* (2nd ed., pp. 581–607). Cambridge University Press.

Callahan, P. (2000). Inter-subjective qualitative landscape interpretation: A contributing research methodology in the exploration of the "edge city." *Landscape Journal, 19*(1–2), 103–110. https://doi.org/10.3368/lj.19.1-2.103

Capaldi, C. A., Dopko, R. L., & Zelenski, J. M. (2014). The relationship between nature connectedness and happiness: A meta-analysis. *Frontiers in Psychology, 5,* 976.

Chen, F., & Thwaites, K. (2013). Chinese urban design. The Typomorphological Approach. Surrey, England.

Chen, Z., Schulz, S., He, X., & Chen, Y. (2016). A pilot experiment on affective multiple biosensory mapping for possible application to visual resource analysis and smart urban landscape design. REAL CORP 2016, Hamburg, Germany.

Creswell, J. W. (1998). *Qualitative inquiry and research design: Choosing among five traditions.* Sage Publications.

Deming, M. E., & Swaffield, S. R. (2011). *Landscape architecture research: Inquiry, strategy, design.* Wiley.

Florian, L., Peter, K., Leila, H., Fabian, S., Heike, T., Philipp, S., & Michael, D. (2011). City living and urban upbringing affect neural social stress processing in humans. *Nature, 474*(7352), 498–501.

Groat, L. N., & Wang, D. (2002). *Architectural research methods.* J. Wiley.

Haluza, D., Schönbauer, R., & Cervinka, R. (2014). Green perspectives for public health: A narrative review on the physiological effects of experiencing outdoor nature. *International Journal of Environmental Research & Public Health, 11*(5), 5445–5461.

Hartig, T., Korpela, K., Evans, G. W., & Gärling, T. (1997). A measure of restorative quality in environments. *Scandinavian Housing and Planning Research, 14*(4), 175–194.

Hassenpflug, D. (2013). Der urbane Code Chinas. Basel.

Kwan, M.-P. (2007). Affecting geospatial technologies: Toward a feminist politics of emotion. *The Professional Geographer 59*(1), 22–34.

Lafortezza, R., Carrus, G., Sanesi, G., & Davies, C. (2009). Benefits and well-being perceived by people visiting green spaces in periods of heat stress. *Urban Forestry & Urban Greening, 8*(2), 97–108. https://doi.org/10.1016/j.ufug.2009.02.003

Lang, P. J., Greenwald, M. K., Bradley, M. M., & Hamm, A. O. (1993). Looking at pictures: Affective, facial, visceral, and behavioral reactions. *Psychophysiology, 30*, 261–261.

Lang, P. J. (1995). The emotion probe. Studies of motivation and attention. [Research Support, U.S. Gov't, P.H.S.]. *American Psychologist, 50*(5), 372–385

Maas, J., Verheij, R. A., Groenewegen, P. P., de Vries, S., & Spreeuwenberg, P. (2006). Green space, urbanity, and health: How strong is the relation? *Journal of Epidemiology and Community Health, 60*(7), 587–592. https://doi.org/10.1136/jech.2005.043125

Mauss, I. B., & Robinson, M. D. (2009). Measures of emotion: A review. *Cognition and Emotion 23*(2), 209–237.

Roe, J. J., Aspinall, P. A., Mavros, P., & Coyne, R. (2013). Engageing the brain: The impact of natural versus urban scenes using novel EEG methods in an experimental setting. *Environmental Sciences, 1*(2), 93–104.

Seresinhe, C. I., Preis, T., & Moat, H. S. (2015). Quantifying the impact of scenic environments on health. *Scientific Reports, 5*, 16899. https://doi.org/10.1038/srep16899

Tost, H., Champagne, F. A., & Meyer-Lindenberg, A. (2015). Environmental influence in the brain, human welfare and mental health. *Nature Neuroscience 18*(10), 1421–1431.

Tzoulas, K., Korpela, K., Venn, S., Yi-Pelkonen, V., Kaźmierczak, A., Niemela, J., & James, P. (2007). Promoting ecosystem and human health in urban areas using green infrastructure: A literature review. *Landscape and Urban Planning, 81*(3), 167–178.

Velarde, M. D., Fry, G., & Tveit, M. (2007). Health effects of viewing landscapes–Landscape types in environmental psychology. *Urban Forestry & Urban Greening, 6*(4), 199–212.

Wang, M., & Ruan, T. (2015). Conflicts in the urban renewal of the historic preservation area— Based on the investigation of Nanbuting community in Nanjing. In Q. Pan, & Cao, J. (Eds.), *Recent developments in Chinese urban planning.* Selected Papers from the 8th International Association for China Planning Conference, Guangzhou, China (pp. 99–120). Springer.

Wang, P., & Meng, Q. (2015). Analysis about key indicators of high density blocks in the new central district from the perspective of mitigating urban heat island: A study at Guangzhou. In Q. Pan, J. Cao (Eds.), *Recent developments in Chinese urban planning.* Selected Papers from the 8th International Association for China Planning Conference, Guangzhou, China (pp. 11–30). Springer.

Wang, X., Rodiek, S., Wu, C., Chen, Y., Li, Y. (2016). Stress recovery and restorative effects of viewing different urban park scenes in Shanghai, China. *Urban Forestry & Urban Greening, 15*, 112–122. https://doi.org/10.1016/j.ufug.2015.12.003

Sebastian Schulz studied Geography and Urban & Regional Planning Management. He has worked in Germany as an urban and transport planner and conducted several planning and consultancy projects in Germany and in the Middle East. Between 2016 and 2020, he has worked for different companies on sustainable planning, mobility and transport planning and eco-city development projects in China. He also actively lectures on sustainable urbanism and smart city development and has participated in several international conferences. Between 2013 and 2019, Sebastian Schulz has worked as a Research Advisor at Tongji University, College of Architecture and Urban Planning (CAUP) in Shanghai. Through this role, he has supported several research projects in the fields of urban landscape planning and sustainable development, as well as several Sino-German joint projects related to research and education. Since 2021, he works in Germany as a transport planner.

Zheng Chen is currently an Associate Professor in Landscape Architecture at Tongji University in Shanghai. She holds a bachelor's degree in landscape architecture, a master's degree in urban planning from Tongji University and a Doctorate degree in Architecture and Design Research from Virginia Tech. She is presently researching the environmental cognition and its application to enhancing well-being by the means of built environmental design. She has conducted research studies to evaluate environmental emotional responses and well-being, by measuring physiological responses via wearable biosensors, including EEG (Electroencephalography), skin conductance, EKG (Electrocardiography), EMG (Electromyography) and other biosensors. Furthermore, she also explores the potential of integrating such real-time bio-signals via HCI (human-computer interface). Two of her research projects have been awarded funding by the Chinese National Science Foundation as well as the Shanghai government.

Chapter 21
Active Ageing—Research on Creating a "Barrier-Free" Outdoor Living Environment for Home-Based Residential Care

Ping Zhang, Chengfang Wang, Siqi Yin, and Shenmao Yang

21.1 Introduction

In recent years, the elderly population in China has grown rapidly. To respond to issues such as changes in the demographic structure, the decrease in family sise and growing number of "empty-nest" families, which result in the weakening of the ability to provide care within the family and a series of other problems, the State has proposed a "9073" Elderly Plan[1] (Fig. 21.1), where it puts forward home-based residential care as the main care model for the elderly promoted by the government and society. Residential areas take on the main burden of elderly home-based residential care and as such, the configuration of their physical environment directly affects whether the elderly live comfortably and happily in their later years (Farquhar, 1995) However, provisions in the existing physical environment generally do not match the basic needs of the elderly which is a problem that has led to a growing discrepancy, especially among the low-income elderly living in affordable housing on the outskirts of urban areas.

Consequently, based on an analysis of the subjective needs and daily lifestyles of the elderly, this paper explores correlations between the basic living environment in

---

[1] The "9073" Elderly Plan refers to 90 per cent of the elderly population being cared for within the family, 7 per cent in community home-based care services, and 3 per cent in institutional care.

In this paper, residential areas refer to the residential compound or community in which the elderly reside. They typically contain residential housing and surrounding facilities such as shops, banks and other amenities. Home-based residential care refers to the elderly living in their own home within a residential compound and receiving care within the community.

P. Zhang (✉) · C. Wang · S. Yin · S. Yang
School of Architecture & Art Design of Hebei University of Technology, Beichen District, China
e-mail: zhangping139@126.com

© The Author(s) 2025
Deutsche Gesellschaft für Internationale Zusammenarbeit (GIZ) GmbH et al., (eds.),
Sustainable Aging, https://doi.org/10.1007/978-3-662-69139-7_21

Fig. 21.1 "9073" elderly plan (*Source* own graphic based on Zhang et al., 2018)

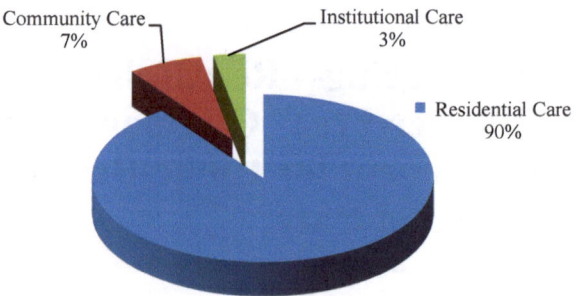

home-based residential care and the needs of the elderly. It proposes an ideal lifestyle model for the elderly, based on the concept of active ageing, so that they can live in truly "barrier-free" residential care in the community. The meaning of "barrier-free" used throughout this paper does not refer to the environment or place in which the elderly live, a standard design based on physical body size and behaviours, nor does it refer to the design of auxiliary services to assist in daily activities; instead it focusses on the daily lives of the elderly in home-based residential care ensuring that they are in their original and familiar environment, meeting both their spiritual needs and physical needs, as well as satisfying their requirements for convenience, comfort, independence and safety. This is precisely what the bundle of basic amenities in residential areas must deliver.

21.2   Analysis of Lifestyle Characteristics of the Elderly and Their Current Living Environment

21.2.1   Analysis of Lifestyle Characteristics of the Elderly

21.2.1.1   Elderly Travel Characteristics

Studies and research into the behaviour of the elderly across different disciplines, show that with an increase in age, the elderly begin to experience a decline in their physical functions. This leads to problems such as legs and feet no longer working as they used to and elderly people becoming easily tired, which in turn results in changes in mobility and the way they travel.

Such characteristics are mainly reflected in three aspects: mode of travel, distance of travel and purpose of travel. Referring to the mode of travel, although per capita car ownership among the elderly has increased, and widespread preferential public transport policies have provided many possibilities for travel, the elderly's primary consideration for themselves is their mobility and road safety. Among the multiple modes of travel available, the elderly mainly opt for walking. In terms of distance to travel, the space in which the elderly are active is different from that of young people's space. The latter focus on their home and workplace forming an elliptical shape of travel activity. For the elderly on the other hand, their home is the centre of all their activities, forming a circular shape (Huang & Wu, 2015) of travel activity, with a significantly shorter distance to travel, mainly concentrated within a 1 km radius. Regarding the purpose of travel, however, there is a shift in the focus of the elderly from survival to lifestyle, whereby the purpose of daily travel is mainly for shopping, leisure and entertainment (Mei, 2008).

The daily life of the elderly is thus mainly concentrated within a 1 km radius of their residential area. According to their average walking speed (0.8 m/s) and a tolerance level of 10–15 min (Quinn & Stott, 2011), an elderly person will take 10 min to walk about 0.5 km and 15 min to walk about 0.75 km. According to a long-distance tolerance survey of the elderly, a half-kilometre walking distance is manageable for most of them, 0.75 km is tolerable for some of them and a walking distance of one kilometre is manageable for only very few.

21.2.1.2 Characteristics of the Needs of the Elderly

The daily basic lifestyle needs of the elderly can be divided per activity type into leisure, shopping, doctor visits, and so on. Subsequently, residential areas must be equipped with elderly/community activity centres, markets, supermarkets, clinics, and various other types of amenities to meet their diverse needs; all the while taking into consideration the decline of their physical functions, their mobility needs and an optimisation of paths leading to the amenities within the residential environment.

The basic needs of the elderly can be mainly characterised objectively in terms of frequency of use and subjectively in terms of importance. However, due to factors such as location, distance, and security, the relation between frequency of use and needs is undefined, whereas the relation between subjective importance and needs is a positive correlation. There is basically a negative correlation between frequency of use of amenities by the elderly and subjective importance and needs. For instance, activity centres, markets and parks which feature high up in the needs of the elderly do not have a high frequency of use, but it is commonly agreed that these amenities are very important; the frequency of use of basic amenities such as clinics, restaurants etc., is essentially 1–2 times per week; meanwhile, the elderly tend to visit banks, post offices etc., 1.5–3 times per month, proving a lower demand and importance attached to these amenities than the first two types. There is a correlation between the frequency of use of paths to reach amenities and the distance, accessibility, landscaping and gardens, road aesthetics and other aspects. Among them, the elderly are mostly concerned about factors such as distance, road safety, and barrier-free design, which should become the main focus regarding the design of the outdoor path environment (Quinn & Stott, 2011; Shin et al., 2011).

21.2.2 Analysis of the Current State of Basic Amenities for the Elderly

Considering the behavioural trends and needs of the elderly, the authors propose an improved potential model for the reasonable allocation of amenities. Using ArcGIS to map out and obtain statistics on 12 basic daily service amenities in four residential areas in and outside the second ring road in Shijiazhuang city, including markets, shops, supermarkets, clinics, pharmacies, restaurants, banks, post offices, bus stops, subway stations, elderly/community activity centres, and parks; then applying a model formula to calculate a reasonable configuration, the results show, firstly, the distance to the amenities is not practical. The most needed amenities such as markets, elderly/community activity centres and parks have poor accessibility and are situated beyond the optimal distance the elderly can walk. However, most stores, banks, restaurants and clinics/pharmacies are within acceptable limits for the elderly. Secondly, there is a lack of variety of amenities. Of the 8 surveyed residential areas, all had some of the "12 amenities" missing. On average, each residential area had about 6 or 7 amenities, but there is a serious lack of markets, parks, and squares. Thirdly, the condition of amenities in residential areas vary considerably. The quantity, type, and accessibility of the amenities in the four residential areas within the second ring road are better than those in residential areas situated outside the second ring road. Moreover, it appears that in the overall urban area, as the peripheralisation of residential areas intensifies, the location of amenities gathers towards the residential areas, but the type and quantity of the amenities show a tendency to decrease (Zhang et al., 2018). Therefore, in affordable housing areas, there are significant problems in the establishment of basic amenities for the elderly, including unreasonable distances to reach them, lack of variety, and fewer options. Furthermore, the quality of life within the second ring road is better than that outside.

21.3 Building an Ideal Model for an Active Elderly Care Life

The transformation of the elderly community in terms of age, physical ability, family environment, economic ability and social status means that the ability of the elderly to face various environmental changes and take on risks is far lower than that of young people. This kind of change significantly impacts the elderly in enjoying equal rights to life. But rights affirm and interpret people's subjective values and are a substantial element for the survival and development of modern society today. In the process of building a socialist and harmonious society, the equality of rights is the equal distribution of resources and a respect for human dignity and basic human rights (Mei, 2008) whereby, it is only by compensating the interests of vulnerable groups that overall equality in society can be achieved in a more just way. Therefore, it is necessary to meet the basic needs of the elderly, to enable this vulnerable group to enjoy true equality of rights and ideal living conditions for active elderly care. This is especially true for the elderly living in residential areas in the outskirts of cities that currently cannot enjoy a good quality of life due to the poor quality of their environmental and a low-income capacity.

This research, based on the principles of health, participation and security, carries out a classification and integration of elderly daily behavioural activities and basic amenities, to create an ideal lifestyle model for active elderly care which solves the problems of security, convenience and comfort and other problems faced by the elderly in daily life due to a lack of amenities.

21.3.1 Principles Behind the Creation of the Model

Being active and healthy in old age means that the elderly remain active and healthy both mentally and physically. To improve elderly people's quality of life, at the same time as satisfying their basic lifestyle amenity needs, elderly participation and integration into society needs to be advocated, which requires the support of the government and society to ensure they get their right to life. Therefore, a model must be created based on the principles of participation, entertainment, support, care and learning (Table 21.1), which proposes the residential area as the main activity centre for the elderly in home-based residential care, and at the same time, fulfils an active ideal lifestyle for the elderly to enjoy self-worth, leisure, medical care and lifelong learning.

21.3.2 Model Creation System

Health, participation, and security are key pillars for an ideal lifestyle model in active ageing care. Based on the "five principles" for the elderly, health and safety complement each other, improving the basic infrastructure for the elderly to travel and carry out activities, and providing favourable conditions for the social participation of the elderly so that they can participate in different activities while gaining physical and mental health.

The concepts of health, participation and security are all defined at the residential area level, through analysis of elderly behaviour. The research seeks to identify any correlations between health, participation, security and behavioural trends and amenities (Fig. 21.2).

Health—To ensure the basic needs of the elderly are met, equal life rights are enjoyed and their daily life activities are facilitated. This includes meeting medical, food and travel requirements through the provision of amenities such as clinics, restaurants, supermarkets, and bus stations.

Participation—To enable the elderly to participate in social activities consistent with their own abilities, needs, and interests, to gain social recognition, rediscover

Table 21.1 Ideal lifestyle model

Design principle	Concept	How it is reflected in the ideal lifestyle model
Participation	Refers to the elderly, in post-retirement, using their own knowledge, techniques and experience accumulated over the years to continue to make new contributions	Social participation of the elderly includes joining in volunteer service organisations and contributing to social and cultural heritage, via actions such as Elderly Volunteer Centres
Entertainment	Refers to carrying out cultural and sports activities suitable for the elderly, so that they can spend their late years with dignity	Satisfying a series of behavioural needs by participating in entertainment, social communication, and interaction in places such as parks, squares etc.
Support	Refers to the elderly having all the support they require in their late years, mainly from their children	On this basis and considering special family structures with the elderly living alone, the model joins together relevant institutions to solve the lack of support for the elderly, e.g. elderly care centres
Care	Refers to the elderly being able to count on family and society's help in case of being unable to personally resolve life matters	While obtaining relevant support, there is also medical care available to help the elderly solve problems such as seeking medical treatment and purchasing medicines, e.g. clinics and pharmacies
Learning	Refers to the mastering of new techniques and new knowledge by the elderly in line with their own hobbies or society's needs, not only uplifting the mind and the spirit, but also learning a new skill via social participation	Realises the goals of participation through learning and coming together through learning by encouraging the elderly to think and try for themselves, and actively initiating participation in organised activities

their own values, and improve self-awareness. This mainly includes the provision of amenities to meet their recreational needs such as parks, elderly centres and voluntary organisations providing social services, etc.

Security—To enable the elderly to obtain supplementary care for their basic needs through the community and via other means, to reduce their sense of loneliness and helplessness and to fulfil their social dependence needs, by mainly including facilities for providing care and on-site services, such as elderly care centres etc.

21.3.2.1 Diversification of Amenity Types

The current lack of variety of amenities in residential areas for home-based residential care, will not be changed by merely designing residential areas in accordance

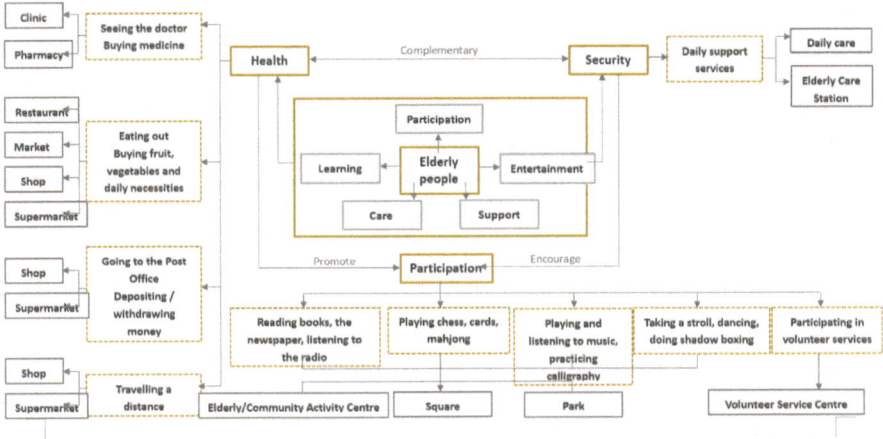

Fig. 21.2 Ideal model for home-based residential care life (*Source* Zhang et al., 2018)

with the standards and norms of ordinary amenity requirements. To begin with, allocation of amenities should be based on the "ideal lifestyle model for residential care" within the concept of active ageing. The variety and types of amenities should be increased to meet the subjective needs of the elderly and to offer a more objective and comprehensively rich offering in residential areas. The result ensures more diverse amenities providing basic services in residential areas.

The positioning of basic amenities in the residential area under the home-based residential care model should be within the maximum daily manageable walking distance (1 km) for the elderly. These should be amenities that firstly satisfy the basic needs of the elderly—medical, food and travel—where they can access timely help in case outside care is required, to ensure their health and safety.

Secondly, improving the quality of life of the elderly by allowing them to participate in social activities, promoting a change in the "quality" of their lives, thus achieving the "five principles" of an ideal lifestyle in care.

21.3.2.2 Hierarchical Layout of Amenities

The key to realising a rational layout of the residential area is to focus on a human-centered spatial layout of amenities from the outset. Since the original design and construction of residential areas did not consider the behavioural characteristics, capabilities and needs of the elderly from the get-go, the result is that the elderly living in home-based care nowadays encounter many inconveniences despite living in a familiar environment, and their quality of life has declined rather than improved.

Layout design adapted to the elderly should fully consider the needs of the elderly in terms of convenience. Drawing on previous research and analysis, one should first examine the distance the elderly walk; a one-kilometre walk is the furthest that the elderly can endure, 0.75 km is tolerable for some, and a walking distance of half a kilometre is feasible by almost all. Therefore, the activities of the elderly should be divided and defined on a standard scale of 0–0.5–0.75–1 km, so that the basic amenities required by the elderly in the residential area are clearly defined and positioned accordingly. Secondly, streetscapes and aesthetics are also factors of concern for the elderly when they are out and about, especially the positioning of shade and rest areas. Rest areas can be positioned in line with the walking distance the elderly can endure, whilst shade areas are a continuous space, and can be composed of trees and other structures.

21.3.2.3 Road Safety

The environmental spaces suitable for the elderly in residential areas should meet their requirements for convenience, safety and comfort. After mobility, safety is the second most important factor affecting the elderly in their activities. In spatial design, the number of intersections and crossings in one path must be reduced. For the elderly living in residential care in the community, they mostly get around by walking, thus moving relatively slowly, however the rapid development of urban areas has seen roads widened and traffic congestion increase, both affecting the safety and easy with which the elderly reach their destinations. Therefore, while distance travelled and transport choice are important factors, planners must also consider path and pavement conditions, road inclination, the decipherability of pedestrian traffic lights, and the composition of the route (Van Cauwenberg et al., 2012). As well as improving the safety, convenience, comfort, diversity and connectivity of travel routes for the elderly, a "loop" network should be created between the destination and the place of residence, rationally positioning basic amenities along the path, in line with the needs of the elderly. Furthermore, there is a need for barrier-free facilities in the outdoor environment, sensibly designed road ramps, pavements, and more barrier-free signs and speed bumps to enhance the travel choices of the elderly and improve the quality of their outings.

21.3.2.4 Standardising to Match the Environment

The amenities currently established in residential areas are often rather simple and rudimentary and cannot reasonably and effectively guarantee the basic living conditions of the elderly. This makes day-to-day outings to go shopping or take part in leisure activities extremely inconvenient and means that the physical and psychological needs of the elderly are not satisfied.

Furthermore, the level of basic amenities in residential areas varies greatly across different locations. An evaluation of the basic amenities in residential areas suitable for the elderly is needed, in line with the ideal active elderly care model, as well as targeted solutions to address the differences in residential areas, to better guarantee "barrier-free" access for the elderly in home-based residential care.

For residential areas to be suitably designed and adapted for the elderly, they must first take into account the special features of the location, the needs of the elderly and the degree of demand for various basic amenities. This data can then be used to correctly plan the configuration of all the different types of basic amenities, as well as considering influential factors such as distance and quantity. Markets, elderly/community activity centres, parks and other amenities which are in high demand by the elderly, should be placed within a 0.5 km radius of their activity space (i.e. the space within their 0.5 km radius in which they go about their activities such as shopping, walking etc.), while clinics, pharmacies, restaurants and other amenities, which are in low demand, should be placed within a 0.5–0.75 km radius. Amenities such as banks and post offices meanwhile, which have an obviously lower demand than the former two types, should be placed within a 0.75–1 km radius. Secondly, since the elderly tend to visit multiple destinations when they venture out, the accumulation of different types of amenities should also be considered when designing the layout. When the elderly go out, their behavioural needs, such as purchasing daily necessities, fruit and vegetables, playing cards or chess, must be satisfied. Moreover, a further configuration according to the specific relationship between the market, the store, the elderly/community activity centre and other amenities will help to improve service ability. Finally, one must also consider the establishment of amenities between adjacent residential areas, and set up, as much as possible, the residential compound as a centre to intensify and share resources.

21.4 Conclusion

This chapter, based on the analysis of the subjective needs and daily lifestyles of the elderly, explores the composition of the outdoor space environment for home-based elderly care—the relevance between basic amenities and the needs of the elderly, targeting the three big principles in active ageing which are "Health", "Security" and "Participation", to create the ideal lifestyle model for the elderly. Constructing an environmental system suitable for the elderly to live according to amenity types, layout, accessibility and matching criteria, and promoting the establishment of an environment in residential areas that is adapted to the elderly, will truly produce a "barrier-free" life in home-based elderly care.

References

Farquhar, M. (1995). Elderly people's definitions of quality of life. *Social Science & Medicine, 41*(10), 1439–1446. https://doi.org/10.1016/0277-9536(95)00117-p

Huang, J. Z., & Wu, M. (2015). Analysis of the travel characteristics of the elderly in megacities and related factors—Taking Shanghai city centre as an example. *Urban Planning Forum, 2*, 93–101. https://doi.org/10.16361/j.upf.201502012

Mei, P. (2008). Ethical thinking on the equal rights of the harmonious society. *Jianghuai Forum, 227*(1), 21–25. https://doi.org/10.3969/j.issn.1001-862X.2008.01.004

Quinn, T. J., & Stott, D. J. (2011). Functional assessment in older people. *British Medical Journal, 343*, 346–352. https://doi.org/10.1136/bmj.d4681

Shin, W. H., Kweon, B. S., & Shin, W. J. (2011). The distance effects of environmental variables on older African American women's physical activity in Texas. *Landscape and Urban Planning, 103*(2), 217–229.

Van Cauwenberg, J., Clarys, P., & De Bourdeaudhuij, I. et al. (2012). Physical environmental factors related to walking and cycling in older adults: the Belgian ageing studies. *BMC Public Health, 12*(142). https://doi.org/10.1186/1471-2458-12-142. Retrieved from: https://bmcpublichealth.biomedcentral.com/articles/10.1186/1471-2458-12-142#citeas. May 24, 2021.

Zhang, P., Yin, S. Q., Wang, C. F., & Shu, P. (2018). A study on demand-supply matching degree of basic living service facilities for the elderly in affordable apartment districts on the metropolitan fringe—A case study of Shijiazhuang. *Architectural Journal, 2*, 95–99. https://doi.org/10.3969/j.issn.0529-1399.2018.02.017

Ping Zhang Doctor of Architectural Design and Theory of Tianjin University, Associate Professor of Architecture and Art Design College of Hebei University of Technology, Master's Supervisor, Deputy Director of Healthy Settlements Research Center, State-level Registered Architect, Member of China Architectural Society, and one of the initiators of "Beijing-Tianjin-Hebei Health Industry Technology Co-innovation Center". Mainly engaged in the construction of the environment and elderly health, medical architectural design, the elderly living environment, public housing design teaching and research. Presided over and participated in more than 20 national and provincial scientific research projects, among which presided over the completion of the national natural science foundation project "Urban low-income elderly public housing endowment strategy and mode research—Taking Tianjin as an example", and published many research papers in the top Chinese professional journals of Architectural Journal.

Chengfang Wang Master of Art and Design, Hebei University of Technology. The research direction is healthy landscape design and its theoretical research. Graduated from Hebei University of Technology and awarded Excellent Graduation Design in 2017 with the title of "Tianjin Huayuan Bihuali community suitable for ageing landscape design". As the person in charge, has completed the National College Students Challenge Cup Competition and the College Students Innovation and Entrepreneurship Competition.

Siqi Yin Master of Urban and Rural Planning, Hebei University of Technology. The research direction is human settlements environment and sustainable development. Graduated from Hebei University of Technology, awarded the excellent graduation project of 2017 at the school level, entitled "General Planning and Design of the New Campus of Chengdu Institute of Physical Education". The University Students' innovation and entrepreneurship competition was awarded provincial and ministerial prises and published research papers in the top Chinese professional journals of Journal Architecture.

Shenmao Yang Doctor of Architectural Design and Theory of Tianjin University, Professor of Environmental and Architectural Art College of Tianjin Academy of Fine Arts, Master's Tutor, National First-Class Registered Architect and Member of China Architectural Society, has accumulated abundant research in the aspects of age-friendly living environment, product line platform construction, general design and campus architectural design. Has presided over and participated in many planning and construction projects, including the Zidu Shanghai Wafer C District Project, which won the Gold Award of the National Habitat Classic Architectural Planning and Design

Competition in 2009, and the Shanghai Architectural Society Architectural Creation Award in 2009.

Chapter 22
Approaches and Guiding Principles for Age-Appropriate Neighbourhood Development in Rural Areas—Experiences from Germany

Kathleen Schmidt and Frank Schwartze

22.1 Introduction

German society is ageing. On the one hand, the process of demographic ageing includes the quantitative increase in the number of older people in society. The reasons for this are the higher life expectancy of each individual and the steady decline in birth rates (Friedrich & Schlömer, 2013). On the other hand, there is a qualitative change in age structures, i.e. the process of ageing is more diverse than it was twenty years ago.

The increasing diversity in old age is reflected in the changed values and attitudes of older people. Thus, the choice of residential location is no longer only explained by the phase of life and the relocation behaviour is no longer only explained by a changed household size. In addition, the demands on everyday life have been adapted to different lifestyles (cf. Kramer, 2007: 401ff.; Poddig, 2006: 214f).

Secondly, social and technological change have an influence to the extent that changed family and household forms as well as more diverse living and employment biographies of people become visible into old age. Since it is closely linked, the use of a smartphone and the Internet in everyday life will become a normal part of modern life into old age.

The change in social conditions is thus changing the structures of the older age groups as a whole and, at the same time, the coexistence and life of each individual. The ageing of society thus presents politicians and planners with major challenges with applied research being in demand. The changed structures of old age and the individual changes associated with ageing are examined empirically in an interdisciplinary manner and recommendations for action are formulated (cf. Gerhards, 2017;

K. Schmidt (✉) · F. Schwartze
TH Lübeck, University of Applied Sciences, Lübeck, Germany
e-mail: kathleen.schmidt@th-luebeck.de

© The Author(s) 2025
Deutsche Gesellschaft für Internationale Zusammenarbeit (GIZ) GmbH et al., (eds.),
Sustainable Aging, https://doi.org/10.1007/978-3-662-69139-7_22

Rüßler, 2015). The findings find their application in the planning of social security systems, socio-spatial planning and in neighbourhood development.

The following article deals with an age-appropriate neighbourhood development in Schleswig–Holstein. The aim is to derive recommendations for action on the basis of empirically gained knowledge in three communities in rural areas for an age-appropriate neighbourhood development oriented towards the approach of healthy living and living up to an old age.

Schleswig–Holstein is the northernmost federal state of the Federal Republic of Germany. Parts of the federal state are in the metropolitan region of Hamburg, which is characterised by dynamic growth processes, but large parts are traditionally characterised by rural areas. The focus of the study is on everyday life in rural areas. The explanations are based on the findings from three teaching research projects which have been carried out since 2016 at the Technical University of Lübeck in the field of civil engineering by the Urban Planning and Urban Development department together with master students in architecture and urban planning and development.

22.2 Demographic Ageing in Germany and Schleswig–Holstein

The demographic and social changes in Germany have led to a change in population and age structures. The age structure of the population will change fundamentally. Currently, the proportion of young people up to the age of 20 is roughly 19% and those in the age group of 65 and older is 21% (Statistisches Bundesamt, 2016: 10). The demographic development, however, is causing a shift in the situation. For example, the 13th coordinated population forecast for Germany shows that the proportion of older people aged 65 and over will rise to around 28% by 2030 and to 33% by 2060 (Statistisches Bundesamt, 2017). Overall, the elderly population group will therefore increasingly influence decision making in politics and planning. These include the future organisation of social security systems, the provision of housing and goods and services, and the establishment of social and medical infrastructure.

The statistical surveys conducted by the federal government, the states and municipalities[1] provide quantitative information on current population trends and structures, by age group, and for various spatial reference levels. Thus, social and socio-spatial planning and neighbourhood development is giving important clues as to which structural-spatial and socio-spatial structures are necessary and will become so in the coming years.

---

[1] The municipality is a political and administrative unit with its own territory. It is a territorial body and forms the lowest level in the administrative structure of the Federal Republic of Germany. The municipalities differ according to the degree of their independence and size into municipalities belonging to districts, cities belonging to districts, cities independent of districts and municipalities integrated into administrative communities (ARL 2010).

The qualitative changes in age structures are reflected, among other things, in changes in norms and values. This is accompanied by the changed role of the family and a change in employment (Krack-Roberg et al., 2016; Schäfers, 2012: 97).

Thus, the generation of parents of today's older people lived in a traditional family focused role. This initially included support for the children and grandchildren. When age-related restrictions occurred, these were then broken down for the elderly within the framework of a so-called generation contract. These life plans are changing due to singularisation and heterogenisation. Family life and the role of grandparents is changing due to a lack of children on the one hand, and due to the increasing geographical distance to adult children on the other. When family support can no longer be guaranteed, the demands on existing care and provision concepts change.

The phase of life at an advanced age usually includes the transition to retirement between the ages of 55 and 69. This is when working life ends and a new phase of life begins. Due to the increase in life expectancy, a further span of 15–20 years can be assumed (Statistisches Bundesamt, 2018: 13). This is largely an active phase characterised by social participation and mobility in space and time and associated with certain demands on the living environment.

A look at the regions in Germany shows that demographic ageing varies from region to region, so that there are areas that are more strongly affected than others (Wiest et al., 2015: 19). On the one hand, this development is influenced by the departure of younger population groups and by the remaining older people. On the other hand, there are regions that are particularly attractive for older people to spend the last phase of their lives in. A large part of Eastern Germany includes regions from which young people tend to leave. The regions that experience an immigration of older people include above all areas that are scenic and interesting for tourism, such as coastal and alpine regions.

The eastern coastal areas in Schleswig–Holstein belong to the areas of migration of older people. At the same time, the proportion of over 60-year-olds in the entire federal state is already higher than that of under 20-year-olds. Current population forecasts calculate that the proportion of people aged 60 and over will be 36% of the federal state's total population by 2030 (Landesportal Schleswig–Holstein, 2018). Here again are regional differences that lead to different strategies and concepts.

22.3 Action-Oriented Research Projects on Age-Appropriate Neighbourhood Development

In the Faculty of Civil Engineering at the Technical University of Lübeck, action-oriented research projects for age-appropriate neighbourhood development are being carried out during the semester. Methods of applied research are combined with practical knowledge transfer in the master courses of architecture and urban planning. In the past three years, three projects were carried out, each of which was focused on a typical Schleswig–Holstein municipality in rural areas for a period of six months.

The necessity of such research and its transfer into teaching and counselling of municipalities is a task for the future to meet the challenges of an ageing population in the municipalities. The municipalities in rural areas that were investigated within the framework of the educational research projects are part of the rural area in Schleswig–Holstein, which is affected by demographic ageing. In addition to overall urban planning, ideas are needed here, especially on the neighbourhood level, which will enable people to live up to an old age.

The municipality has the role of mediator and taking over the findings of the studies carried out and the development of conceptual ideas in its planning and political decisions. On the one hand, the municipalities are a specific planning case in teaching and at the same time practice for future planners. On the other hand, these three case studies are representative of many rural areas in Schleswig–Holstein which face major challenges due to demographic ageing.

Each of the three projects made it possible to access the demands on the residential location in old age. Depending on the structural, spatial and socio-spatial framework conditions, the fields of action in which the municipality can become active were identified. Under the heading "Healthy living and living into old age", the existing structures were examined, with the needs of an ageing population in the foreground. The knowledge gained in this way led to conceptual ideas and concrete project proposals which later could be implemented.

In the three municipalities examined here, three existing quarters were the focus of interest. In other words, the research concentrated on the existing structures of the housing estates, on the socio-spatial interrelationships and on the perspectives of the people who live their everyday lives in the neighbourhoods. In each project, the strengths and challenges of urban development were recognised and discussed and conceptually worked on against the background of spatial structural change, demographic development and the demands of healthy living in the community up to a ripe old age.

The projects dealt with the effects of the ageing population on the community and how living and living space should be designed so that the demands of older people are considered. To this end, the structural, spatial and socio-spatial structures on site were first analysed, using different survey and evaluation methods.

Firstly, the methods included the preparation of the overarching and municipal planning within the framework of a literature analysis with the aim of recording the development intentions of the planners and political decision-makers. Secondly, the available statistical data were evaluated within the framework of a secondary data analysis with the aim of summarising the population and social structures, the building, open space and supply structures as well as the traffic infrastructure in text and cartographically form preparing them. Thirdly and additionally, discussions were held on site with randomly selected persons living in the municipalities. In addition, two events took place in each of the three projects, one at the beginning and one at the end, to which the mayor, the local council, the responsible planners and district managers as well as interested residents were invited for an open exchange about the projects.

The first step of the analysis was followed by step two, the development of conceptual ideas, which, depending on the size of the research area, were combined into a neighbourhood development concept. This also included the development of specific project proposals for the implementation of the concept.

In the following, the three projects are presented which exemplify the development of everyday life in old age in rural areas in Schleswig–Holstein. These include.

- The critical view of living in an institutional form of living, such as a nursing home, using the example of Lensahn, with a focus on neighbourhood development at the location of the nursing home
- The investigation of age-appropriate neighbourhood development in an existing neighbourhood, using the example of Lauenburg, with a focus on the development of the neighbourhood at the location of the the Arbeiterwohlfahrt (AWO[2]) facilities.
- The development of alternative forms of housing with a focus on private domesticity using the example of the municipality of Nusse and the aim of developing a town for healthy living into old age.

22.4   Out of the Nursing Home into Life in the Quarter in Lensahn

The municipality of Lensahn is located in the district of Ostholstein in Schleswig–Holstein. In a scenic and attractive location on the Baltic Sea there is a direct connection to the A1 motorway. Due to its convenient location, Lensahn assumes central supply functions for the 5011 inhabitants (as of 31.12.2017) and for the inhabitants of the surrounding communities (Statistikamt Nord, 2017). Lensahn is a popular place to live and relax. In addition to the social facilities for children and young people, the AWO is the provider for institutional living in old people's and nursing homes as well as private apartments for the elderly in Lensahn.

In Schleswig–Holstein, the AWO operates around 200 facilities of this kind with a total of 3500 employees and 2600 volunteers. At the Lensahn location, the AWO offers full inpatient nursing care in the nursing home with the "Haus am Mühlenteich". In addition, the household and nursing infrastructure of the care facility are used to be active in outpatient care. For example, people who cannot or do not want to take care of themselves at noon can order lunch from the "Haus am Mühlenteich", which is then delivered to their homes.

The community of Lensahn was selected as a teaching research project because it represents all the communities in Schleswig–Holstein in which the AWO provides

---

[2] The Arbeiterwohlfahrt is a non-profit organisation and is one of the six leading associations of voluntary welfare work in Germany. It maintains more than 13,000 facilities and services in all federal states, such as homes and flat-sharing communities, day-care centres for children and young people and the elderly, information and counselling centres, outpatient services, including social welfare services, counselling centres, day-care centres and workshops.

inpatient care for people of older age. The full in-patient care facility in Lensahn is home to people who have generally reached the age of 80 and for family or health reasons cannot or do not wish to remain in private homes. The relocation takes place by liquidating the private household with selected personal items included in the move. People who are usually familiar with their surroundings, because they have lived in Lensahn or in one of the surrounding communities and have grown old, live in the facility.

It turned out that the demands of living in the house at the Mühlenteich are changing. Thus, it is no longer only the good supply of regular meals and the possibility of nursing services that are in the centre of interest. Rather, wishes were expressed for alternative accommodation locally, in the community of Lensahn, the relationship between living in the house at the Mühlenteich and in the surrounding neighbourhood or a place of exchange and social interaction.

These changes were the starting point for the teaching research project during the winter semester 2016/2017. The focus of the study was the development of a structural and socio-spatial concept for an age-appropriate neighbourhood development. The project involved students of the masters' course in architecture, the head of the Arbeiterwohlfahrt Pflege Schleswig–Holstein, the management and employees of the house at the Mühlenteich, the commune with the mayor, the head of the Lensahn Ordnungs- und Planungsamt and residents of the house and the community of Lensahn.

In a first work package, the wishes and ideas of older people about living in Lensahn at an old age were questioned. For this purpose, the students of the Master's programme in Architecture polled 45 people aged 57 and 85 based on a partially standardised questionnaire about their current living situation, everyday life in their homes and surroundings as well as everyday places in the community.

The evaluation of the survey led to the second work package, the formulation of a spatial concept for the conversion and new construction of the AWO site as well as to a list of criteria for the age-appropriate district development around the Mühlenteich. In this way, the classic building design was combined with the requirements of an age-appropriate neighbourhood development. The students planned a new location that would be green, airy, barrier-free and diverse, combining institutional and new lifestyle in old age.

It turned out that the new building of the nursing home at the same location corresponds to the wish to remain in the community of Lensahn. The establishment of a day care facility offers interested parties an alternative to moving out and in. The concept of a flat-sharing community for people with dementia is an important facility to create a suitable framework for this clinical picture. The variety of offers also includes the construction of serviced apartments, which are rented in the same way as living in private homes, but which also offer help in the household and care services. For social interaction, a district café is to be set up and the sensory garden maintained and expanded. The connection between the location at the "Haus am Mühlenteich" and the neighbourhood should be of a structural nature, such as the creation of connecting roads from the buildings and lines of sight out of the building, and of a socio-spatial nature, such as organising afternoon meetings in the district café.

22.5   Neighbourhood Project "Lauenburg-Mitte"

The municipality of Lauenburg an der Elbe has a population of 11,485 (as at 31.12.2017) (Statistikamt Nord, 2017). The hillside location results in a topographic structure of the urban area, which, however, is also reflected in the structural-spatial and current social-spatial structures. The old town, with its castle and historic residential and commercial buildings, is a tourist attraction. It lies directly on the banks of the Elbe and is also known as the lower town. The upper town can be reached on foot from the lower town or by public transport and via two main roads. It is a residential area, centre of supply with food and goods and services for daily needs as well as serving social needs. The Lauenburg-Mitte neighbourhood stretches from the geographical centre to the eastern administrative border. In 2016, a neighbourhood management system was set up for its age-appropriate development. Compared to the city, an above-average number of older people live in this area. Many smaller cities in Schleswig–Holstein face great challenges due to ageing and financial shortfalls. Lauenburg is representative of many other municipalities. It has been a so-called "Consolidation" municipality since 2013, i.e. the municipality can no longer independently finance its administrative responsibilities due to insufficient income and is subsidised by the State.

Three factors influenced the choice of the municipality as a case study for the teaching research project in the summer semester of 2017. The first of these was the special urban planning situation in which the topographical division into an upper and a lower part of the city is accompanied by a separation in terms of building culture and social space. Secondly, with the initiation of a neighbourhood management that is striving for age-appropriate development, an instrument is applied that seems to meet the demands of an ageing urban population. Thirdly, the municipality's financial situation provides, on the one hand, limited scope for urban planning and development and, on the other hand, great challenges in the face of continuing demographic ageing.

In this educational research project, the focus was on an age-appropriate neighbourhood development in the context of overall urban planning that succeeds in incorporating all three of these factors. This called for concrete project proposals on the neighbourhood level that would make it possible to live to an old age. The associated task included conceptual work in the existing quarter, so that here too, as before in Lensahn, the everyday life of the local people could be described and analysed. In the following, conceptual ideas for the quarter in respect of housing, social affairs as well as help and care were in demand. For the actual implementation, existing cooperation on site should be considered to be able to make specific recommendations to the respective parties in the neighbourhood management.

Right from the start, the mayor, the urban planning office and the neighbourhood management of the AWO were available as contacts for the situation on site. During the educational research project, the mayor was able to provide insight into political decisions, the city planning office provided important planning information and the neighbourhood management was an important contact for the needs of the people in the neighbourhood. The students of the master's programmes in architecture and urban planning involved in the project were thus able to discuss their own research and analyses with those responsible in the city of Lauenburg. As was already the case in Lensahn, the political and planning levels were involved in the development of a neighbourhood development concept.

The project in Lauenburg was also carried out in two work packages. First, the current situation was analysed in a first work package, whereby from the very beginning the existing age-appropriate development of the neighbourhood was compared with the planning framework and the urban development situation. This included an analysis of the strengths and weaknesses of the urban landscape, the development and open space structures, local public transport and the situation of motorised individual transport as well as the supply of goods and services for daily needs, and social and cultural infrastructure. The analysis included discussions on site, the evaluation of provided materials and a mapping of Lauenburg-Mitte.

In the second work package, concrete measures and projects were developed in addition to a mission statement and suitable targets for the age-appropriate development of the neighbourhood. Two of the three selection criteria mentioned above for the municipality of Lauenburg were taken up again as a case study.

The division of the city into an upper and a lower urban area in terms of building culture and social space due to the topography is a challenge that needs to be solved, especially for older people. This includes making it possible to reach the city on foot without barriers. Accessibility was interpreted in terms of construction from social and mental aspects. To change the structural situation, the design of the public spaces according to DIN 18040-1[3] and the public transport and open spaces according to DIN 18040-3 is recommended. Here all requirements for age-appropriate building are formulated, which give information and orientation to the architect and the urban planner. The project ideas include, among other things, the design of pavements and the provision of additional furniture, such as seating along typical paths in Lauenburg. The dismantling of social barriers involves creating leisure facilities that are accessible to every purse. Regarding mental barriers, the project addressed the fact that there are areas in the upper city, in the transition from the upper city to the lower city and in the lower city that are referred to as spaces of fear by older people. A specific proposal, better illumination and exact route guidance were worked out.

The neighbourhood management of the AWO is working on an age-appropriate development in the Lauenburg-Mitte neighbourhood. It is part of the strategy to

---

[3] The German Institute for Standardisation e.V. (DIN) has defined requirements for barrier-free building. DIN 18040-1 defines the barrier-free planning, execution and equipment of publicly accessible buildings and their outdoor facilities. The aim of DIN 18040-3 is the accessibility of buildings so that they can be accessed and used by people with disabilities in the usual manner, without particular difficulty and in principle without outside help (according to § 4 BGG Behindertengleichstellungsgesetz).

diversify and enlarge the services already offered into the neighbourhood with additional services for older people. The focus here is on improving the quality of life, opportunities for participation and the care situation of older people.

The teaching research project follows on from this to devote itself to neighbourhood management as an instrument that looks at the demands of an ageing population and summarises them in a development concept. For this purpose, all conceptual ideas were combined into a total of three goals:

Objective 1: A volunteer neighbourhood aid has been set up and is used by the residents of the neighbourhood. This objective takes up the changes in the population and social structure by making the immediate neighbourhood increasingly important when everyday help is needed and the family members do not live in the immediate vicinity due to work obligations. This includes, among other things, the use of smartphones and the Internet to organise various everyday activities, such as shopping or meeting like-minded people. A contact person and discussion times in the neighbourhood café have been defined obligatorily. For the coming years, it has been worked out that walks should be organised on a regular basis so that conversations can be held on the way about everyday life and problems addressed. In addition to the health-promoting activity, it is continuous communication that makes neighbourly help possible, whereby the development of a neighbourhood platform on the Internet was also recommended, which, for example, provides information about events in the neighbourhood.

Objective 2: The participation of the residents of the neighbourhood is guaranteed for the entire duration of the project. The aim is to ensure that the residents participate in the development of the neighbourhood. One project aimed at achieving this goal is the neighbourhood newspaper, which provides information about everyday life in the neighbourhood and important dates. For further participation, the network should be expanded to include volunteer caretakers. These are people who live in the neighbourhood, who keep an eye on the development in the neighbourhood and give the local people everyday assistance.

Objective 3: There are opportunities for older people to meet in the neighbourhood. There are opportunities for age-appropriate exercise and leisure activities. To achieve this goal, the Quartiers café in the residential- and service centre was opened and has been the central meeting point for older people in the neighbourhood ever since. In addition, at the time of the teaching research project, it was established that projects such as an encounter and movement park, a senior theatre and a multi-generation garden would also contribute to preventive health promotion.

22.6   Alternative Living in the Town Centre and in the Resthof in Nusse

The municipality of Nusse is in the district of Herzogtum-Lauenburg and belongs to the district of Sandesneben-Nusse. Nusse has 1.091 inhabitants (31.12.2017) (Statistikamt Nord, 2017). It is a place with a historical settlement core between Lake Nussen and Lake Ritzerau. The former settlement has developed into a regional centre from where supplies can be sourced. Nusse has a good infrastructure. There is a pharmacy, a general practitioner's practice, a paediatrician and a dentist for medical care. There is also a midwifery practice, occupational therapy, physiotherapy, speech therapy and a veterinary practice.

The municipality of Nusse was chosen for the teaching research project because, as one of many rural municipalities, it wants to face the process of ageing and the associated planning challenges for neighbourhood development and the handling of the historical stock of buildings worthy of preservation, the so-called residual farms. The project investigated which structures should be kept in nuts and created for a healthy everyday life into old age. With the settlement structure, the supply and the creation of alternative forms of living in the Resthof, three fields of action were identified to be worked on for Nusse within the framework of an age-appropriate neighbourhood development.

The special feature of the project in Nusse was the initiative by the local council to approach the authors of this contribution with the wish to investigate and conceptually summarise the development of alternative forms of housing for the entire village within the framework of a teaching research project. Thus, the mayor and the members of the municipal council were available from the beginning as contact persons, who are responsible for the planning decisions, like the development of residential land and industrial areas in the municipality.

From the analysis of the current situation and on the basis of the discussions with community representatives, it was possible to work out that the currently existing village centre should be preserved as a centre and at the same time the release of further residential land should be limited to dense development. The analysis shows that the existing infrastructure will be within walking distance and within easy reach, regardless of age, using a compact development approach. In this context, the routing within the administrative boundaries of the municipality has also been examined and considered worth improving. On the one hand, the main roads are assessed as having low barriers. On the other hand, paths linking the main roads must be created. With this high-density area development and enhancement to the pathways, it will be possible to reach not only the supply points but also the recreation areas such as Lake Nussen.

The proposals for the development of settlement structures are linked to the goal that all public and private facilities should be accessible up into old age. Firstly, this includes the creation of barrier-free public paths, buildings and squares. Secondly, there is the area of health-promoting measures with the aim of promoting mobility. Among the specific project proposals for the town centre is the introduction of an active neighbourhood aid system and the construction of a playground for young and old.

The area of action for the remaining farms is linked to the two objectives of preserving the formative settlement structures and creating suitable forms of housing for old age. A residual farm is a structurally preserved unit that once belonged to an agricultural enterprise. The remaining farmhouses in Schleswig–Holstein are predominantly of the indoor hall building type, characterised by a large living area with five or more rooms and are mainly inhabited by older people. In Nusse, the remaining farms are located along the main roads and closer to the town centre than to the municipal boundaries.

People of older age still wish to remain in private domesticity and in the familiar living environment, even if age-related restrictions occur. Within the framework of the educational research project, it became apparent that three areas of action are connected with this wish in relation to the Resthof: Firstly, the creation of an age-appropriate apartment with barrier-free access; secondly, assistance and supply at the residential location from the neighbourhood and customary services; and thirdly, the connection of the Resthof to the supply facilities, the recreation areas, the cultural, social and sports facilities of the community as well as the local public transport system for trips beyond the community borders into the region.

As alternative forms of housing for the elderly in Nusse, the educational research project was able to show that there is interest in communal housing projects, regardless of whether they are self-administered or offered by an organisation such as Arbeiterwohlfahrt. Everyday life in the community is conceivable as both homogenous for the elderly and heterogeneous for the generations to come. The living space and the number of rooms in the remaining courtyards make it possible for three to five people to live together in one apartment or house.

Nursing and medical care at the residential location is to be organised on an outpatient basis. There are already care service providers who are active in the public health area. This offer must be maintained. In addition, attention should be paid to the possibility of providing goods and services for daily needs, for example in the form of delivery services or home visits. As was already the case in Lauenburg, it was worked out for the municipality of Nusse that an Internet platform, on which on the one hand information is provided about the various areas and on the other hand an exchange between the residents of the municipality can take place, not only addressing the use of smartphones and the Internet in everyday life, but also satisfying the need for information and exchange across all ages.

22.7   Guiding Principles for Age-Appropriate Neighbourhood Development

The three teaching research projects show that in rural areas, such as those examined here, closed local structures and the mixture of functions are important when healthy living and housing up to old age is one of the envisaged development goals. In addition, attention must be paid to the accessibility on foot of selected service areas, which, however, also differ depending on the initial conditions, be it socio-spatial or and/or structural-spatial. Ultimately, due to the increasing life expectancy and thus prolonged and active life phase of older age, there is an urgent need to pay attention to the preservation of spatial mobility.

It should also be noted that rural areas are characterised by traditional family structures and that the development of alternative forms of everyday life have so far only been marginal. However, as the three projects show, the use of new technology, the computer, the Internet and smartphones is constantly increasing in everyday life and not only changes the user behaviour of people, but also implies changed demands on the residential location, as long as local digitalisation is available.

Based on the above-mentioned points, which were to be established within the framework of the synthesis, in the following paragraphs, work action packages were developed for the three research areas, which are always structured in the same way. They begin with the presentation of a mission statement for the site, the neighbourhood or the location and the associated objectives and measures leading to specific project recommendations.

A model for neighbourhood development is a projection into the future, which contains objectives for the development of the study area based on the analysis and evaluation of the existing situation.

In the rural areas, guiding principles were developed which tie in with the strengths and opportunities, but at the same time do not lose sight of the weaknesses and threats. In addition, current models of urban development are included so that concepts can be developed that can be connected. Examples are given:

- The place of short distances—mobile, healthy and networked.
- We bring you together—care accessible to all

In the formulation of the associated objectives and measures, attention was paid to the temporal framework, because this not only makes prioritisation possible, but also evaluation and monitoring of success.

22.8 The Place of Short Distances—Mobile, Healthy and Networked

The concepts of short distances[4] and mobility are to be understood in this model both spatially, in the sense of unrestricted movement in physical-geographical as well as virtual space. This refers to two different areas of teaching research projects. One is the change in the age structure and the functional concept of age, with which the statement is connected that the phase of life of people of higher age means an independent, active time lasting 15–20 years on average. On the other hand, from the conceptual planning to application becomes clear: it is one of the declared goals of the state government in Schleswig–Holstein to make rural areas attractive for all age groups by means of suitable mobility concepts in both real and virtual space.

Thus, as shown by the example of Lauenburg, one of the short-term goals can be to initially strengthen non-motorised individual transportation. On the other hand, in the medium term local public transport will be adapted to the actually desired frequency of the resident population. And in the long term, it is feasible that a small city like Lauenburg can develop into a Smart City, for rural areas more appropriately referred to as a Smart Village. Here, too, the focus is on increasing attractiveness for different target groups. The slow, age-related restrictions at the end of the life can be shaped by a future-oriented neighbourhood development in terms of construction, space, social space and technology.

22.9 We Bring You Together—One Supply that is Accessible to All

The concept of short distances and mobility is closely linked to the model of care and services accessible to all up to old age, whereby a selected functional area is dealt with in greater detail here. The underlying thesis is that if care is assured and made accessible to all, exodus can be avoided and the population in rural areas stabilised. This applies to all ages, but especially to the elderly, in which a local supply within walking distance and a barrier-free access offer age-specific service. This not only provides the possibility of active ageing, but also concretely prevents a single elderly person from moving from private domesticity to an institutionalised form of living.

Using the example of Lauenburg and Nusse, it was established that a healthy life on site is firstly supported by the availability of services within walking distance and barrier-free accessibility, and secondly that the supplies are expanded through access to digital providers. Closely linked to this is the objective of expanding the range of driving services. Here, too, the long-term perspective is to counter age-related restrictions by creating suitable structures in the town or neighbourhood.

---

[4] This is the application of the concept of the city of short distances (Umweltbundesamt 2011).

22.10 Recommendations for Action for Municipalities in the Rural Areas of Schleswig–Holstein

Five recommendations for action will be developed for the municipalities in rural areas in Schleswig–Holstein: The independent life phase, the mixture of functions, the alternative forms of housing, the regional differences and digitisation.

Firstly, any spatial planning should consider the fact that higher age comprises an independent life phase of 15–20 years, which is becoming increasingly diverse due to the ongoing changes in population and age structure. The phase of life is predominantly actively experienced. From the age of 80 onwards, age-related restrictions occur, so that everyday activities are increasingly oriented towards the place of residence.

Secondly, it is essential to secure or create function-mixed spaces for everyday life, considering the user's perspective. Here it is significant which population structures currently live on site and how they are going to develop over the next few years. If, for example, it is a question of an age-homogeneous population structure, certain demands will be placed on the residential location. One possibility for empirical investigation of user perspectives is offered by action space research (Nash, 2015; Oswald & Konopik, 2015; Pelizäus-Hoffmeister, 2014; Wilde, 2014), which focuses on activities outside the home and investigates the paths and places of everyday life.

As a third recommendation for action, the creation of alternative forms of housing in towns and neighbourhoods is formulated. These forms of housing serve, as it were, the increasing diversity of the demands made by older people on their residential locations. The creation of a broad range of different forms of living makes it more likely that they will remain within their familiar surroundings.

The fourth recommendation is derived from the experience of teaching research projects that regional differences must be considered in every concept development. With the use of thorough status analyses, appropriate survey methods and a summary evaluation, a suitable concept can be developed.

Fifthly, the digitisation of rural areas contributes to enabling people to live a healthy and long life and to live well into old age. The group of older people has also arrived in the digital age. The computer, the Internet and a smartphone are part of their everyday lives. Especially where there is a lack of public and private facilities, ordering via the Internet and using delivery services are important alternative to organising everyday life.

22.11 Knowledge Transfer–Lessons Learnt

The evaluation and results of the three exemplary teaching research projects show that the design and development of age-appropriate places and neighbourhoods fit into the canon in terms of content and process and the requirements of sustainable urban development. It becomes clear that it cannot be the task, apart from very specific approaches for people with certain limitations such as dementia, to design urban neighbourhoods in such a way that they meet the needs of only one age group quantitatively or qualitatively. The aim must be to achieve a diversity and openness of spatial and structural structures as well as of infrastructural options and services through inclusive planning that is open to all in the spirit of universal design.

The implementation of this goal focuses more on the planning process and less on the outcome, the result of planning, being a sustainable, humane city for all groups and age groups as shown above. The experiences from Germany and the described study projects illustrate the shift from a policy and planning for certain groups or requirements to that of an open and inclusive planning policy.

The increasing diversity of age groups, their demands and wishes as well as the limits of economic viability have led in Germany to a growing questioning of the concepts based on the care concept of an independent old-age and nursing infrastructure separate from the other structures. So far, however, there has been little or insufficient information and knowledge exchange about and between the different actors and institutions, that would have to be involved in the joint planning to develop a city for all. The results of the three projects presented here clearly demonstrate how interdisciplinary the approaches to solutions range from architectural culture to mobility.

Bringing together the previously separate worlds, with which separate financial support systems, legal requirements and professional responsibilities etc., is therefore a new and major challenge, especially for the municipalities. In this context, the practice-oriented teaching research projects of the university not only serve to enhance the qualification of architecture and urban planning education for the needs of ageing cities, but as real laboratories offer municipalities the opportunity to identify future tasks and make them comprehensible.

Two aspects can be derived from the example for the challenges for the cities resulting from the rapid demographic and social change and the ageing population in China. Like other areas in which China has skipped certain spatial development patterns that have shaped Germany in the last five decades there is the chance to skip the concept of building an infrastructure exclusively geared to old-age provision and to focus primarily on the development of the required cities for all. This implies, however, as a second aspect, that the demands and needs of an ageing population must be included more offensively and comprehensively in the planning and design of cities and neighbourhoods. One way to achieve this is to expand the educational content of central occupational fields such as architecture, urban planning and social work. In both countries there is the possibility of joint learning about specific related challenges.

References

ARL, Akademie für Raumentwicklung in der Leibniz-Gemeinschaft. (2010). Gemeinde/Kommune. Retrieved from: https://arl-net.de/en/lexica/de/gemeinde-kommune, May 01, 2019.

Friedrich, K., & Schlömer, C. (2013). Demographischer Wandel. Zur erstaunlich späten Konjunktur eines lang bekannten Phänomens. *Geographische Rundschau, 1*, 50–56.

Gerhards, P. (2017). Nachbarschaftsbeziehungen älterer Menschen - Subjektive Konzepte und Hilfe-potenziale. Eine Untersuchung organisierter und nichtorganisierter Nachbarschaft, Technische Universität Kaiserslautern. Dissertation

Krack-Roberg, E., Rübenach, S., Sommer, B., Weinmann, J. (Bundeszentrale für Politische Bildung (BpB), Eds.), (2016). Formen des Zusammenlebens. Retrieved from: http://www.bpb.de/nachsc hlagen/datenreport-2016/225884/formen-des-zusammenlebens, March 31, 2018.

Kramer, C. (2007). Alt werden und jung bleiben. Die Region München als Lebensmittelpunkt zukünftiger Senioren? *Raumforschung und Raumordnung, 65*(5), 393–406

Landesportal Schleswig-Holstein. (Ed.), (2018). Demographischer Wandel - Bevölkerungsentwick-lung bis 2030. Retrieved from: https://www.schleswig-holstein.de/. November 10, 2018.

Nash, C. (2015). Veränderungen des Raum-Zeit-Verhaltens im Zuge von Lebensumbrüchen und ihre Anforderungen an die Stadt- und Verkehrsplanung am Beispiel des Eintritts in den Ruhestand, Universität Kassel. Dissertation

Oswald, F., & Konopik, N. (2015). Bedeutung von außerhäuslichen Aktivitäten, Nachbarschaft und Stadtteilidentifikation für das Wohlbefinden im Alter. *Zeitschrift Für Gerontologie Und Geriatrie, 48*(5), 401–407.

Pelizäus-Hoffmeister, H. (2014). Gesellschaftliche Teilhabe Älterer durch Alltagsmobilität. *Forum Qualitative Sozialforschung FQS, 15*(1), 23.

Poddig, B. (2006). Die „Neuen Alten" im Wohnungsmarkt. Aktuelle Forschungsergebnisse über eine stark wachsende Zielgruppe. *vhw Forum Wohneigentum, 3*, 211–217.

Rüßler, H. (2015). *Lebensqualität im Wohnquartier. Ein Beitrag zur Gestaltung alternder Stadtge-sellschaften* (1st ed.). Kohlhammer.

Schäfers, B. (2012). *Sozialstruktur und sozialer Wandel in Deutschland* (9th ed.). UTB.

Statistikamt Nord. (Eds.), (2017). Bevölkerung der Gemeinden in Schleswig-Holstein. Retrieved from: https://www.statistik-nord.de/fileadmin/Dokumente/Statistische_Berichte/bevoelkerung/ A_I_2_S/A_I_2_vj_174_Zensus_SH.xlsx. September 30, 2019.

Statistisches Bundesamt. (Ed.), (2016). Ältere Menschen in Deutschland und der EU, Wies-baden. Retrieved from: https://www.destatis.de/DE/Themen/Gesellschaft-Umwelt/Bevoelker ung/Bevoelkerungsstand/Publikationen/Downloads-Bevoelkerungsstand/broschuere-aeltere-menschen-0010020169004.pdf;jsessionid=2F6162DD94DE0B20EDADDF8A3605A174.liv e712?__blob=publicationFile. July 5, 2019.

Statistisches Bundesamt. (Ed.), (2017). Bevölkerungsentwicklung bis 2060. Ergebnisse der 13. koordinierten Bevölkerungsvorausberechnung. Aktualisierte Rechnung auf Basis 2015. Retrieved from: https://service.destatis.de/bevoelkerungsyramide/#!y=2060. November 10, 2018.

Statistisches Bundesamt. (Eds.), (2018). Sterbetafel 2015/2017. Methoden- und Ergebnisbericht zur laufenden Berechnung von Periodensterbetafeln für Deutschland und die Bundesländer. Retrieved from: https://www.destatis.de/DE/Publikationen/Thematisch/Bevoelkerung/Bevoel kerungsbewegung/PeriodensterbetafelErlaeuterung5126203177004.pdf?_blob=publicationF ile. November 10, 2018.

Umweltbundesamt. (Ed.), (2011). Leitkonzept - Stadt und Region der kurzen Wege. Gutachten im Kontext der Biodiversitätsstrategie. Retrieved from: http://www.uba.de/uba-info-medien/4151. html. August 09, 2019.

Wiest, M., Nowossadeck, S., & Tesch-Römer, C. (2015). Regionale Unterschiede in den Lebenssi-tuationen älterer Menschen in Deutschland (Deutsches Zentrum für Alternsforschung (DZA), Ed.), (DZA-Diskussionspapier No. 57), Berlin. Retrieved from: https://www.dza.de/fileadmin/ dza/pdf/Diskussionspapier_Nr_57.pdf. September 15, 2018.

Wilde, M. (2014). Mobilität und Alltag. Einblicke in die Mobilitätspraxis Älterer Menschen auf dem Land. Friedrich-Schiller-Universität Jena, Dissertation, Wiesbaden.

Kathleen Schmidt has worked since 2011 as a Research Associate for local and urban development at the TH Lübeck, University of Applied Sciences, Germany. She studied Geography at the Technical University of Dresden with a focus on urban and regional planning. After gaining her diploma in 2006, Kathleen worked in regional management at private institute for communicative urban and regional planning and research (KoRiS Hannover). Between 2008 and 2010, she was a scientific associate of the research group for forecasts and urban development of the Leibniz Institute of Ecological Urban and Regional Development (IÖR Dresden). Her doctoral thesis investigated local life contexts of older people aged 55–69 years. Her main areas of interest and research are healthy urban and local development for an ageing population. She is particularly interested in the transfer of action-oriented research into teaching and the use of empirical methods in planning practice.

Frank Schwartze has worked as a professor for Urbanism and Planning at the TH Lübeck, University of Applied Science, Germany. He studied Urban and Regional Planning in Berlin and Venice and obtained a postgraduate Diploma from the University of Rouen, France. In 1997, he joined the Brandenburg University of Technology (BTU) as assistant professor in the department of Urban Design. From 2009 to 2013, He served as the s head of the BTU's Department of Urban Planning. As an expert for urban planning and development, Frank has been engaged in several development cooperation projects of UN Habitat and GIZ in East and South-East-Europe and the MENA-Region. His main field of expertise is related to forms and processes of sustainable urban development. He is specifically interested in strategic planning, instruments and tools for the steering of urban development and urban regeneration processes.

Chapter 23
Design Strategies for Elderly Care Communities in a Diverse Society

Yiqun Guan and Yiming Liu

23.1 The Impact of a Diverse Society on the Elderly Community

23.1.1 Diversification of Family Structures and Filial Piety

In the past, people did not move around as much, and the elderly usually chose to spend their later years in their home environment. Traditional family values still exist today, but new opportunities brought about by the reform and opening, and empty nest families that came along with urbanisation, have opened cracks into the present culture. Young people move far away from home, whilst the elderly stay behind, causing cracks in community life in the neighbourhood, and shrinking family structures. At the same time, many new generations are beginning to choose a new way of life, living in a DINK (dual income no kids) household and without marrying, as managing individual finances becomes fashionable, the family structure is diversified.

Accompanying all this comes a shift in values. Where once people returned to their roots and the family's children took care of their parents, the elderly now tend to leave their family home to enter an elderly care facility, which has created large-scale elderly care communities. When families evolve into smaller units, the breakup of the well-established pattern makes such values and the practice of filial piety no longer the only course (Fei 2004). Providing for your parents as they once did for you is no longer the only ethical choice. The elderly care community has given a new significance and many new ways to support the relationship between young people and the elderly.

Y. Guan (✉) · Y. Liu
Shanghai Qicheng Construction Planning & Design Company Ltd, Shanghai, China
e-mail: yiqun.guan@gn-int.com

© The Author(s) 2025

Deutsche Gesellschaft für Internationale Zusammenarbeit (GIZ) GmbH et al., (eds.),
Sustainable Aging, https://doi.org/10.1007/978-3-662-69139-7_23

23.1.2   Diversification of Lifestyle and Consumption

Shift in Consumer Concepts

Over the past few decades, the economy has driven the transformation of values and the shaping of consumer societies. Credit consumption, mature commercial insurance, sound security systems, and the rise of new media have all impacted the way people spend their money, and have widened the differences in consumer concepts between generations.

At the same time, as the average level of education rises, the ability of the elderly to make independent choices increases. From a business perspective, employee health and retirement are also part of a company's strategy; revenue growth has meant that current users of elderly care services and those who will be eligible for such services in the next 50 years have very different levels of disposable income. As divorce rates and bachelor rates are rising, the current generation of 20-somethings are already preparing financially for their retirement, with health and medical care becoming one of the most important drains of funds. In the future, the elderly will no longer continue to wait passively, but will engage in active consumption.

Diverse Lifestyles

At present, the elderly have many diversified lifestyle choices such as tourism and meditation, and they have far more possibilities than before. Choices are no longer limited to one world and elderly lifestyles will be more diverse in the coming decades. The elderly who have experienced new opportunities in the millennium, have gone from leading a simple life focusing solely on taking care of their families after retirement, to a varied life of self-fulfilment and contributing to society (Murata 2015).

Furthermore, with the wave of Internet of Things and Artificial Intelligence, intelligent homes and wearable devices have become popular, catering for customised elderly lifestyles. Retirement in the future will feature big data and smart systems, and this trend will also provide a variety of new ideas for the elderly.

Individual Differences Within the Same Generation

At present, daily life in the post-industrial era is full of stress, the gap between the rich and the poor has widened, the crisis of confidence intensifies, and behind their electronic screens individuals have increasingly significant differences. Since the media has provided more opportunities for people to express themselves, values have diversified, and people have begun to embrace difference (Cheng and Simmel 1999). Different individuals within the same generation also choose significantly different ways of caring for the elderly. Within the same generation of people, some plan to benefit from specific elderly care services, some return to the countryside to settle (urban "hollowing out"), some tenaciously opt for eating healthily and exercising into their later years. This multi-level approach is derived from the choices of today and is quite different from the singular option the elderly had in the past. The provision of elderly care services should also follow and constantly adapt to diversification. When the generation born after the reform and opening begin to enter elderly life, it will be even more diverse.

23.1.3 Regional Differences

Urban–Rural Dual Structure

Since the rapid development of urbanisation, the changes brought about by the urban–rural dual structure have offered a multi-level approach to providing for the elderly in urban and rural areas. In the countryside, there are still many old people who maintain the respect for ritual customs by the land. Due to a relatively backward productivity level, they have tasted the bitterness of life, and their needs in later life are simpler and more intensive. The loss of labour and the lack of corresponding elderly care facilities are also problems that need to be urgently solved in rural areas.

Those who have experienced the popularisation of education and flocked to the cities in the wave of reform, benefitting from a relatively high economic level, will explore more in life and during retirement, being able to combine a way of life and values from the countryside with their urban lifestyle, thus producing different wants and needs. But how to reasonably accept the elderly and face the problem of ageing in a large city where the population begins to overflow is also a challenge.

Other Regional Differences

Regional differences exist not only within the urban–rural structure, but also in the demographic structure, population flow, economic development level, and urbanisation process of different provinces. There are differences in the structure and values of the labour force, in the ability to provide elderly care services, and in the consumption capacity of the elderly population. This uneven level of development has created a diverse range of elderly communities. For example, a well-equipped elderly community can care for more customers, but in areas where values are not particularly open, where there are profoundly traditional concepts, and the family unit tends to consist of many children, limiting the outflow of the population, is a further challenge for institutional elderly care.

23.2 Design Strategies for Elderly Care Communities

Dealing with a highly diverse society, one must think outside the box when designing and constructing elderly care communities, stimulating a more open discussion and cooperation model. The following considerations about design strategy are not only passive responses to the status quo but should also be seen as positive steps for the future.

23.2.1 User Orientation and User Participation

Stakeholders involved in the design process of existing care communities are mainly investors, builders, designers and operators. The concepts of priority for ROI, cost, and efficiency are the dominant factors controlling the overall project development.

For users—the concern of the elderly remains mainly on passive demand satisfaction. The absence of user participation in the project process may result in it lacking perspective and market resilience. Therefore, a design approach based on the user's perspective is particularly important to maintain a design value centered on the user—the elderly. During the design process, one must always maintain the ability to "empathise" with the elderly, and change from being an "observer" to becoming a "participant". Every stakeholder can find the opportune time to conduct road show presentations based on the viewpoint of the elderly, furthermore, inviting elderly representatives to participate in the discussion and even finding solutions that are both creative and effective.

23.2.2   A More "Inclusive" Design

Inclusivity in design can accommodate a wider range of user lifestyle habits. An elderly care community with good barrier-free design is an important basis for the concept of "inclusive" design. But an excellent barrier-free design is not just providing a simple handrail and ramp access, it is firstly a continuous and complete system, and secondly, it also offers multiple preventative solutions to fit different habits and spatial characteristics.

An "inclusive" elderly community moreover needs to give more diversity, flexibility and growth to the living space. The creation of a diverse space environment is the basis for providing more personalised service capabilities, which allows users to have more diverse choices. However, making any early arrangements for the functional space will pose real challenges in the operational phase. There are many examples of wasted or insufficient space, so creating a space that has a certain flexibility or elasticity is of real value. We tend to set up "buffer zones" between different functional spaces, leaving room for growth; or adopting a "universal" design approach to switch between different application scenarios (see Fig. 23.1).

23.2.3   Spatial Environments that Encourage Independent Living

Traditional elderly care services and spatial design focus on "helping", and we advocate that this concept can gradually shift to "supporting". Supportive design theory means encouraging the elderly to complete a movement as independently as possible, going from passive acceptance to a more active and autonomous lifestyle. This inspires us to rethink issues such as the relationship between service space and served space, and the degree of involvement in barrier-free design.

Fig. 23.1 Open and flexible use of public spaces create more opportunities for peer-to-peer communication and joint activities. Foshan "Le Rong" elderly care community (*Source* Shanghai Qicheng Construction Planning & Design Company Ltd.)

Associated with this concept is the "ageing in place" approach, which is also an extension of the home-based residential care system within the elderly care community. It allows the elderly to grow old in as familiar a living environment as possible, reducing unnecessary environmental changes. Therefore, it is necessary to make environmental design more "inclusive", which also poses a new challenge to operational management (see Fig. 23.2).

23.2.4   Spatial Organisation from the "Village" Viewpoint

Current elderly care communities resemble more residential compounds with a closed-management system. Public facilities are all concentrated within the compound and its spatial layout is relatively simple, resulting in a daily life for the elderly spent between two-points: the "club" and "home".

We propose to use the "village" viewpoint to organise the space and function of the community. Villages, especially traditional villages in China, have the following typical features: a clear sense of territory, a relatively vague community boundary, multi-level public space, a spontaneously formed informal interaction space, strong integration with nature, stable neighbourly relations, and a strong collective consciousness, etc. (see Fig. 23.3).

This brings us to rethink completely the spatial structure of the elderly care community. Can we create a richer spatial organisation and provide a richer life experience? Can we separate the service facilities from the clubhouse, offering a

Fig. 23.2 The indoor environment of "de-institutionalised" elderly apartments create a comfortable feeling of home. Foshan "Le Rong" elderly care community (*Source* Shanghai Qicheng Construction Planning & Design Company Ltd.)

more "centralised + dispersed" distribution? Can we redefine the community's physical and social boundaries with its operational boundaries? This approach, which is similar to an urban design perspective, is not only relevant for large communities, but also for the organisation of space within small-scale elderly care institutions.

23.2.5 Geographical Features and Cultural Attributes

The design of geographical features is not only reflected in regional climatic conditions, the surrounding environment, the respect for local construction techniques and materials, but also in the respect for the local way of life. At the same time, for the elderly, this is a continuation of the life experience they are familiar with. We often try to find a typical place and arrangement for social interaction in each project location, for example Guangdong's morning tea culture and Hong Kong style cafes, Chengdu's chess culture and tea houses, the northeast's bath culture and public baths, Shanghai's western social interaction and coffee shops and so on (see Fig. 23.4).

Of course, respect for the religious beliefs of the elderly and design responses is also an important topic. In addition to designing different places for religious gatherings in the community, it is even possible to imagine an elderly community that is specifically targeted at people sharing common religious beliefs. If the design

Fig. 23.3 "Village-oriented" community public service facilities create a more interesting spatial hierarchy. Shanghai East Shanghai Yifuhui Senior Care Community (*Source* Shanghai Qicheng Construction Planning & Design Company Ltd.)

is carried out from the perspective of a group of people sharing common cultural or religious backgrounds, it will surely usher in a more characteristic and innovative form of elderly care community.

23.2.6 Highly Integrated Intelligent Platform

To fully meet the individual needs of the elderly in a highly diversified elderly care community, the traditional management model will surely face unprecedented challenges. How to provide a better service response while effectively controlling overall operating costs? The development of Internet technology, Internet of Things technology and Artificial Intelligence technology has provided us with a new service perspective, and has also triggered our rethinking of design ontology: Is spatial organisation in a highly dispersed state possible? Can there be a synergy between

Fig. 23.4 Reception lobby using old Shanghai cultural elements. Shanghai Pujing Elderly Service Apartments (*Source* Shanghai Qicheng Construction Planning & Design Company Ltd.)

the use of space and time allocation? Can the flexible change of space become "predictable"?

23.3   Conclusion

As we face a time of change, the core values of openness and tolerance that the country advocates will infiltrate into the design of China's elderly care communities of tomorrow. We must also adopt a more inclusive attitude to meet the opportunities and challenges of a diverse society.

References

Cheng, B. Q., & Simmel, J. (1999). *A diagnosis of modernity Hangzhou* (pp. 81–91). Hangzhou University Press.
Fei, X. T. (2004). *Earthbound China*. Beijing Publishing House.

Murata, H. (2015). *Consumer behaviour in super ageing societies* (pp. 3–6). Economic New Wave Press.

Yiqun Guan is a Member of Elderly Service Facilities Association a branch of the China Association for Engineering Construction Standardisation. He also works as a Manager of Shanghai XICHENG architectural Planning & Design Co., Ltd. Master of Architecture, Berlin Industrial University of Germany. He has a master's degree in architecture from the Berlin Industrial University of Germany. Mr. Guan's work focuses on ageing architectural design and provides design services for different developers and institutes. He has completed more than 40 projects on new building and retrofit building related to ageing such as, elderly nursing institutions, elderly apartments, rehabilitation hospital etc. He has participated in numerous conferences, workshops and forums related to the elderly issue where he shared his professional insights. He supported knowledge building activities at Beijing University, Tsinghua University, and Academy of Architecture. He has also organised training sessions and lectures for universities, research institutes and social organisations.

Yiming Liu works as a consultant of Shanghai XICHENG architectural Planning & Design Co., Ltd. She holds a master's degree of science in strategy and management from the London School of Economics and Political Sciences (LSE). She has accumulated rich experience in market research, customer study and project valuation. She has provided consulting service to more than 10 projects related to elderly architecture, involving assisted living apartments, elderly apartments etc. She has worked with different developers and institutes developing new lines of products, while conducting dynamic analysis on business strategies in ageing. She has also carried out indepth interviews and data analysis through field visits in many different cities.

Chapter 24
Layout Designs of Institutional Care Facilities in the US, Germany and Japan—The Potential Impact on the Evolution of China's Elderly Care Facilities

Sha He, Deqing Bu, and Bo Zhang

24.1 Introduction

According to the 2017 Circular of the State Council on Issuing the 13th Five-Year National Plan for Developing Undertakings for the Elderly and Establishing the Elderly Care System, it is predicted that by 2020 China's elderly population of over 60s will increase to about 255 million, equal to 17.8% of the total population; the number of oldest-old will increase to about 29 million while those living alone or in an "empty nest" will increase to about 118 million, with the old-age dependency ratio increasing to around 28%. With the rapid ageing of the population, the cost of social security for the elderly will continue to rise and they, along with the disabled elderly will continue to increase in number, making caring for this part of society an increasingly prominent problem. When comparing elderly institutional care facilities with those of developed countries, China still remains in the primary development phase and needs to learn from experiences of advanced elderly care in developed countries.

China has an extensive territory with a large population; its elderly care industry emerged late and has seen a rather staggered development. Judging by the pattern of development of elderly care facilities, layout design has been mainly determined by the level of economic development and design concepts along with other factors. China's varying levels of economic development, social differences and business culture across its different regions has created multiple forms of layout design for China's elderly care facilities. In contrast, the elderly care industry in developed countries emerged relatively early, and, by conducting comparative analyses, the

S. He (✉) · D. Bu · B. Zhang
School of Architecture and Art, North China University of Technology, Hebei, China
e-mail: 775387890@qq.com

© The Author(s) 2025
Deutsche Gesellschaft für Internationale Zusammenarbeit (GIZ) GmbH et al., (eds.),
Sustainable Aging, https://doi.org/10.1007/978-3-662-69139-7_24

authors have found that the stages of development of layout design for care facilities in developed countries have all followed a similar pattern, eventually coming up with a relatively mature design model. The development of anything needs to follow a certain course, it is not possible to leapfrog over development stages. This paper assesses and summarises the evolution of layout design of institutional care facilities in developed countries to identify a common pattern, and, on this basis, develops a layout design model which can suit the varying levels of economic development across China's different regions, and provide a reference for designers of elderly care facilities.

24.2 Comparative Analysis of Layout Design in Institutional Elderly Care Facilities in the US, Germany and Japan

Layout models in the US have evolved over five key stages. Early elderly care facilities were based on hospital standards with 4-bed rooms, limited services and a public space at the end of the corridor (Quan, 2013); in 1978, a semi-private space was included in designs between a straight corridor and private rooms, while multi-bed rooms were replaced with single or double rooms (Zhou, 2016); in 1987, collective housing was introduced and the prototype of the group form emerged, with dining rooms and bathrooms for shared use (Nelson, 2016); and between 1988 and 2005, elderly care facilities completely transformed into a family group structure, the whole system formed by groups of 11 people, all of them linked by a central service area (Nelson, 2016). From 2006 until today, there has been an emphasis on "human-centered" design, with the addition of a transit zone between semi-private areas and private bedrooms, focusing on privacy within the group structure (Nelson, 2016).

Layout models in Germany have evolved over four key stages. Starting in 1940, Germany built the first generation of care homes mainly to provide housing, and usually designed with a long corridor and a very small public space; between 1960 and 1980 a "medical and housing" combination concept was introduced. These care homes resembled hospitals, with additional public space compared to previous designs; in 1980 a "live-in" elderly care system was advocated, which gradually developed into a group home design structure; and the fourth generation of elderly care home design came about in 1995, when people started to focus on a "family atmosphere", paying attention to safety and the psychological feelings of the elderly, thus bringing about the group structure (Schmieg, 2016).

Layout models in Japan have evolved over four key stages. The first stage was primarily focused on multi-bed rooms with a long corridor and rooms on both sides of the corridor, a collective dining space and a spacious public bathroom (Sima, 2015); the second stage differed from the first with the expansion of atrium space and the addition of double rooms; the third stage showed the beginning of gradual development into a group structure, with a central open patio surrounded by small rooms to the side—mostly single rooms. The fourth stage saw the development of group care—each group not exceeding 10 people—with fixed nursing staff and non-connecting groups, providing a family living environment (Hu, 2017).

Since the US, Germany and Japan transitioned to an ageing society at different periods, and developed at different rates, the evolution in layout designs of their elderly care facilities also emerged at different times. However, looking at their development it is fair to say that these three countries share certain commonalities in the evolution of layout designs of their elderly care facilities (see Table 24.4). Architectural design of elderly care facilities in all three countries first focussed on "basic needs" and only later making them "better quality". Development went from focussing merely on basic behavioural needs to be mindful of the psychological feelings of the residents. By taking into consideration elderly people's needs, reducing distance to services, creating a "homely" familiar environment, reducing the feeling of loneliness and reinforcing the duty of care of the nursing staff, layout designs developed from a basic long corridor with rooms on both sides model to a group space design. Drawing on the different periods of development in layout design in different countries and by observing spatial patterns, it is possible to establish a common set of designs to use as a reference for China.

The US, Germany and Japan have different social backgrounds, geographies and local customs which are reflected in the diverse ways the layout design of their elderly care facilities have evolved. By conducting a longitudinal comparative study, the development of public space has gone through a process starting from non-existence, then beginning from a simple to more complex design, and finally developing into a three-level public space system observed today: namely, transition from public space to semi-public/semi-private spaces and finally to a private space. In the fourth and fifth stages of evolution in the US, a transitional zone has been added between the semi-private areas and the private bedrooms, further segmenting the dynamic areas within the small groups. The "family atmosphere" group design in Germany and Japan focusses more on the psychological feelings of the elderly. Germany directly integrates care homes into the community, with assembly room space open to the public. Elderly people's living rooms are separated from different semi-private spaces with the use of furniture providing a resting place for the elderly and reducing their feelings of loneliness, however this also increases the issue of interference of personal space. In its present stage, Japan focusses on multi-level activity space, which does not exist in the US and German designs. The non-connecting groups are fixed to avoid passage between them, but at the same time the elderly are encouraged to participate in inter-group activities as well as social activities outside the facility to establish a connection between them and the community. Consistent with the scope of common activities, activity space is divided into three levels; the first level is for activities on a facility-wide scale, the second is for smaller semi-public activities on a residential group level, and the third is for semi-private activities on a single residential unit level.

Results: By analysing the afore-mentioned countries' layout designs, one can better understand their 4-stage evolution and corresponding 4-type spatial design patterns: The first stage is known as the "corridor" style, the second stage—the "corridor + public space" style, the third stage—"group" style, and the fourth stage—"multi-group + public space" style. The development of public space has gone through a three-level process starting from nothing, to subsequently convert from a public space to a semi-public/semi-private space and finally into a private space.

24.3 Using Environmental Behaviour Psychology to Analyse the Reasons for Commonalities in the Evolution of Layout Design in the US, Germany and Japan

People's psychological characteristics differ at different stages of life, and attention should be given to the particularities of elderly people's psychological feelings during the design process. Layout design of elderly care facilities must not only satisfy the daily needs of old people but must also focus on their mental health. It is possible to understand the importance of the psychological characteristics and needs of the elderly from the evolution of layout design in each country's institutional care facilities. Below, we analyse the reasons driving development of each country's elderly care facilities from the "corridor" style stage through to the "group" style stage from an environmental behaviour psychology perspective:

24.3.1 Placing Importance on the Creation of Privacy and Personal Space

Due to privacy, the elderly place great importance on personal space and their own domain. When in their personal space they feel relaxed, stress-free, and can carry out activities freely. Once their personal space is violated, negative emotions can easily surface, which does not benefit the elderly mentally or physically. Early care homes which were designed with rooms for four people were more gradually developed into homes with single or double rooms better respecting elderly people's privacy and avoiding interference of personal space, thus offering them a comfortable resting area (Hu, 2017).

24.3.2 Reducing the Sense of Rejection and Increasing Communication

In later life, the elderly are less and less in contact with people and things, moreover their physical degeneration and decline in self-confidence, leads them to gradually stay away from crowds, resulting in the feeling of loneliness. The design of a long corridor with rooms on the side and a lack of public space leaves the elderly without a place to interact, while designing a public space at the end of the corridor makes it too far from their resting rooms (Hu, 2017). The final layout design with a living room near the bedrooms provides the elderly with an area to interact, thus reducing the feeling of loneliness and at the same time increasing interaction with the community (Hu, 2017).

24.3.3   Good Spatial Reasoning

As elderly people suffer physical deterioration, just being in a safe and comfortable environment allows them to live at ease. Reducing the long corridor layout in turn reduces the amount of ground the elderly need to cover, benefitting their memory. It is easy to get lost in long and narrow passageways which can make an elderly person feel anxious and depressed. By dividing the whole space into different groups, the elderly are active just within their group, and are restricted from passing through to other groups. This reduces their overall activity space making it easier for them to familiarise themselves with their surroundings (Hu, 2017).

24.3.4   Importance of Creating a "Neighbourhood Feel" and Increasing a Sense of Belonging

The elderly often develop feelings of loneliness and anxiety especially after leaving their own familiar environment to enter an unfamiliar one. The development from a long corridor style to a group style has helped elderly people to disconnect their environment from that of a hospital setting. Grouping them into groups of 10 people allows them to be more familiar with each other and increases their sense of belonging. The fixed nursing staff is familiar with the needs of the elderly in the group, which improves the efficiency of care giving and instils in the elderly a greater feeling of affection from the staff (Hu, 2017).

24.3.5   Focus on Individual Development

Elderly people like to live in an environment that they are familiar with, so, while moving in, it is important to encourage them to bring their own furniture, bedding and other items to allow them to arrange their furniture in their private room as they like, creating a feeling of being at home, and making them calm and relaxed. At present, all three countries encourage the elderly to bring their own furniture and focus on individual development. Certain care homes in Japan install window displays in each room to let the elderly decorate them by themselves, which not only strengthens spatial cognition, but also helps to familiarise them with their memories, and gives them the space for creativity, which is conducive to maintaining physical and mental health (Hu, 2017).

24.4 The Impact of Institutional Care Facilities in Developed Countries on China

In summary, by analysing the layout designs of institutional care facilities in developed countries it is possible to pick out some specific design features as inspiration for China's elderly care facilities.

24.4.1 Layout Designs of Institutional Care Facilities in Developed Countries and Their Potential Impact on China

(i) While designing the layout of a care institution, more attention should be paid to the emotional needs of the elderly. Where possible, single and double rooms should be increased and multi-person rooms decreased, so as to provide the elderly with more personal space.

(ii) There are two obvious characteristics in geriatric psychology; a sense of attachment and symbiosis. Elderly people still feel attached to society, they long to live together with family members and exist in society just like other people. Elderly care facilities can be opened to the public at different time periods, to introduce a more social atmosphere and avoid dissociation from society, while increasing the space for public activities.

(iii) With the improvement of nursing care, it is even more necessary for elderly care facilities design to transition to a group formation, to create a safe and warm environment for the elderly, and to facilitate care given by the nursing staff.

(iv) Restructuring the long corridor style with rooms on each side and reducing the distance the elderly must cover, facilitates memory and allows the elderly to familiarise themselves with their environment. In China, elderly care design should be more "human-centered", and more mindful of the psychological needs of the elderly.

(v) Arranging small group living spaces for up to 10 people helps the elderly to remember one another, strengthens the "neighbourhood feel" and, at the same time, allows the elderly and caregivers to be more familiar with one another, all the while increasing the efficiency of care.

(vi) Encouraging elderly people to bring their everyday belongings and to decorate their own space by themselves, creates a homely feel, thus satisfying their needs to stay in an environment they like for as long as they want.

(vii) According to the scope of common activities, activity space is divided into three levels; the first level is for activities on a facility-wide scale, the second is for smaller semi-public activities on a residential group level, and the third is semi-private activities on an individual residential unit level. It is possible to add a transitional zone between the semi-private and private spaces within the group structure, increasing the levels of space division, which ensures privacy for the elderly within the group space, and respects everyone's lifestyle habits.

(viii) Setting up an appropriate number of multi-person rooms and using furniture to separate spaces, can help to reduce the feeling of loneliness in the elderly.

24.4.2 Useful Layout Designs for Different Regions Across China

China's economic development has been uneven and inadequate across its different regions, and compared to the US, Japan and Germany, China's ageing process and lifestyle habits are somewhat different. However, it is still possible to draw useful references from the evolution of layout designs of institutional care facilities in these three countries. We should adapt to local circumstances and, based on China's national conditions, explore the different levels of economic development across the country to separately apply one, or several, layout combination models.

(i) The first stage "corridor" style layout design is suitable for elderly care facilities in regions with weak economies. The long corridor with bedrooms on both sides has a similar spatial structure to that of a hospital. Although there is a lack of consideration for the psychological and behavioural needs of the elderly, the space utilisation rate of this layout is the highest, and in such areas, the priority is to provide for crucial "basic needs".

(ii) The second stage "corridor + public space" style layout design is suitable for elderly care facilities in less-developed urban areas—a long corridor with rooms on both sides and a public space located at the end or in the middle. This layout offers appropriate space for common activities and for the elderly to interact within limited economic and spatial constraints.

(iii) The third stage "group model" layout is suitable for elderly care facilities in second-tier cities and medium developed areas—developing from a small group model in the beginning. Groups are connected to each other without obvious division and there is no large public space for interaction between groups. The dining room or public bathrooms can be used to connect each group.

(iv) The fourth stage "multi-group + public space" layout is suitable for elderly care facilities in first-tier cities such as Beijing, Shanghai, Guangzhou and Shenzhen which are considered to be developed economy cities. This design makes it easier for the nursing staff to provide care and gives the elderly a sense of "being at home". A semi-private living space is designed within the group space, which is convenient for the elderly to interact with one another. Multiple groups are connected by a common activity room forming a multi-level space, suitable for the organisation of social activities within the community, and helping to avoid feelings of inferiority that the elderly can suffer from.

The four cases mentioned above are just for general reference. The four different spatial layouts are not necessarily meant to fit four different regions exactly. There may be different organisations and different situations in the same region warranting different needs and layouts. It is thus possible that two or three layout models are required in one region. Every spatial layout should be chosen appropriately according to the specific situation, being mindful of economic conditions, the business situation and health status of the elderly in that area.

24.5 Conclusion

China's ageing process, its distinct culture, lifestyle and traditional customs all differ from those of the US, Germany and Japan, but it is still worth acquiring knowledge about layout designs of institutional elderly care facilities from these three counties. The developments in layout designs of elderly care facilities experienced by these three countries are somewhat different but they have something in common, which is that they all focus their design on the needs of the elderly, placing importance on physical and mental health to avoid psychological problems caused by a disconnection from society. We need to combine the universal principles of advanced elderly care in developed countries with the specific needs of China's elderly care and embark on a path of elderly care with Chinese characteristics.

References

Hu, H. Q. (2017). Japanese utilisation of old buildings to provide elderly care facilities and its impact on China. *City and House Journal, 24*(273), 29–33. CNKI:SUN:CSZZ.0.2017-11-006.

Nelson, G. G. (2016). Household models for nursing home environments. Retrieved from December 12, 2020: www.pioneernetwork.net/wp-content/uploads/2016/10/Household-Models-for-Nursing-Home-Environments-Symposium-Paper.pdf

Quan, X. (2013). New trends of design for the community support for old age and the aged apartment in recent years of America. *Architectural Journal, 03*(535), 81–85. https://doi.org/10.3969/j.issn.0529-1399.2013.03.018

Schmieg, H. P. (2016). The development process and trends of elderly care buildings in Germany. *Chinese Hospital Architecture and Equipment Journal, 1*, 63–65. CNKI:SUN:YYJZ.0.2016-01-023

Sima, L. (2015). Yuimarl tamadaira no mori, Hino, Tokyo, Japan. *World Architecture Journal, 11*(305), 64–69. JournalArticle/5b3c2cc3c095d703c0a4190b.

Zhou, Y. M. (2016). Case study on the renovation of high-rise pension facilities—Montgomery house in Chicago, USA. Retrieved from May 24, 2021: http://blog.sina.com.cn/s/blog_6218cf5 70101socs.html

Sha He graduated from North China University of Technology with a master's degree in architecture, during which her research direction was architecture design for the elderly. Now she is working as an assistant engineer in China Shipbuilding Industry Corporation International Engineering Co., Ltd.

Deqing Bu graduated from the department of architecture, Harbin institute of architecture and engineering with a master's degree. Now he is associate professor and master tutor of architecture department of North China University of Technology. He is the deputy secretary general of the medical and health environment branch of the Chinese medical and health culture association, the director of the cold region architecture academic committee of the Chinese society of architecture, the expert member of the theatre architecture and stage machinery academic committee of the Chinese society of stage art, and the member of the Chinese society of interior design. Long engaged in high-rise hotel design research and elderly architecture design research.

Bo Zhang graduated from Tsinghua University with a PhD in engineering. The North China University of Technology, associate dean of engineering professor, master tutor, urban development and heritage protection research institute, Chinese opera institute visiting professor, committee member of architectural society of China's industrial heritage, the landscape technology, thus, China's stage art association, stage engineering and building research centre, deputy director of the theatre.

Chapter 25
Project Case: Integrated Elderly Care Project in Yichang City

Chenzi Yiyang

Facts

Project period: February 2016–August 2016

Target group: Elderly Care Service of Yichang Municipality, Elderly Residents of Yichang, Hubei Province of China

Commissioner: CDIA Fund, Supported by German Federal Ministry for Economic Cooperation and Development (BMZ), Governments of Austria, Governments of Sweden, Governments of Switzerland

Partners: Asian Development Bank, Yichang Municipal Government, Hubei Provincial Finance Department

Country of project activities: China

25.1 Social Challenge Behind the Project

The city of Yichang has a strong need for Elderly Care (EC) support, yet its infrastructure and services could not meet this demand. In 2015, the elderly ratio of Yichang was 13.7%, which was above the national average. In Yichang's urban centre, this ratio was projected to rise to 26% by 2030. More than half of the elderly lived alone or with a spouse and were subject to insufficient care for daily living requirements. This means they lacked assistance with practical activities such as carrying out household chores and making hospital visits, but they were also short of emotional support. Due to the insufficient public EC service and the high-cost private EC service in

C. Yiyang (✉)
Deutsche Gesellschaft Für Internationale Zusammenarbeit (GIZ) GmbH, Bonn, Germany
e-mail: chenzi.yiyang@gmail.com

© The Author(s) 2025   293
Deutsche Gesellschaft für Internationale Zusammenarbeit (GIZ) GmbH et al., (eds.),
Sustainable Aging, https://doi.org/10.1007/978-3-662-69139-7_25

Yichang, there was a heavy demand to establish an affordable EC services system that could cover home-based, community-based and institutional care which also avoids an over-emphasis on expensive institutional care.

25.2 Project Approach

CDIA is a multilateral project development facility co-managed by the GIZ and Asian Development Bank (ADB). It provides technical assistance to medium-sized cities in project structuring to bring infrastructure project concepts to a pre-feasibility stage, which could be linked to downstream finance from multi-development banks or private financiers.

Through the provision of a Pre-Feasibility Study (PFS) for the Yichang Integrated Elderly Care PPP Pilot Project, CDIA supports the Yichang Municipality in improving its elderly care system. A technical assistance fund of USD 300,000 from CDIA was assigned to develop the PFS. The PFS produced a replicable PPP modality for not only Yichang, but also for Elderly Care in China. In addition, CDIA helped to link the project to downstream financiers to bring it to a loan preparation and implementation stage. In 2018, ADB approved a USD 150 million loan for the implementation of the Hubei Yichang Comprehensive Elderly Care Demonstration Project.

25.3 Value Added

CDIA's main value for the project lies in the preparation of the EC PPP modal tailored for the needs of Yichang and linking the project to downstream financing.

CDIA's Technical Assistance via the PFS includes:

- Determining the viability of private sector involvement in the development of the selected pilot facilities and its integration with the overall Elderly Care System.
- Identifying suitable PPP structures to best utilise the private sector's skillset so that the present and future demands for EC can be met in a satisfactory manner.
- Appraising the financial viability of the alternative PPP structures and enhancing the financial viability by developing innovative ways to improve PPP design, utilise ADB funding and increase revenue generating opportunities for the private sector.

25.4 Lessons Learnt

The main challenge of the project lies in increasing the EC standards of Yichang's services and infrastructures while maintaining acceptable prices for such services. The existing low EC price reflects the service of low standards. For example, the existing facilities show low human resources in quantitative and qualitative terms. The outcome of the service standards versus services tariff indicates the necessity of securing project support from the municipal government. This support could be provided to the project through Construction Subsidy and Operational Subsidy.

25.5 Ideas for the Future

For similar projects, government support is necessary in two forms:

- Indirectly i.e. regulating the service standards to avoid other facilities being significantly cheaper due to a very poor quality of service provision.
- Directly i.e. incentivise the local government to provide EC facilities through minimum guaranteed demand mechanism.

Chenzi Yiyang is an advisor for the Sustainable Infrastructure Project at Deutsche Gesellschaft für Internationale Zusammenarbeit (GIZ). She provides advisory to GIZ's cooperation with the Asian Infrastructure Investment Bank (AIIB) on upstream project preparation. Prior to GIZ, she has worked as an urban policy research analyst for the Natural Resources Defence Council and the organisation for Economic Co-operation and Development, covering topics such as land use efficiency, non-motorised transport, and financing green urban infrastructure. She holds a master's degree in Urban Governance from Sciences Po Paris.

Chapter 26
Conclusion: Lessons Learnt and Future Paths for International Cooperation

Marie Peters and Sabine Porsche

Germany and China face similar demographic challenges in similar intensities; both countries have some of the lowest fertility rates worldwide (DW, 2020; Statista, 2021). These demographic challenges affect all realms of life, challenges that have been addressed by the governments of both countries for many years, or even decades, using different approaches. Regulations and guidelines have been published, addressing different fields on various levels, and solutions have been identified and implemented. During the COVID-19 pandemic, however, it became even more obvious that international efforts undertaken in recent decades to address the ageing challenge do not suffice. It has become evident that a framework which allows people to live an independent life if possible has not been enforced sufficiently, and that pivotal issues, such as the financing the care of increasing numbers of elderly in the future, have not been solved to date.

The present book uses the momentum generated by the pandemic to reinforce discussion and foster collaboration in the field of sustainable ageing. It aims to provide a knowledge source for practitioners and policy makers, as well as academic communities, with a particular focus on the topics relevant for the Sino-German cooperation implemented by Deutsche Gesellschaft für Internationale Zusammenarbeit (GIZ) GmbH on behalf of the German government, namely, long-term care insurance, elderly care education and age-friendly cities and communities.

This book demonstrates that international exchange on policies, good practice and solutions is an essential and valuable undertaking, and one worth striving for as we prepare societies for the future. The following chapter summarises the discussion points in these focus fields, recommends areas of action, elaborates on lessons learnt, provides recommendations for international cooperation and finally concludes with a look to future aspects to be addressed.

M. Peters (✉) · S. Porsche
Deutsche Gesellschaft für Internationale Zusammenarbeit (GIZ) GmbH, Bonn, Germany
e-mail: marie.peters@giz.de

© The Author(s) 2025
Deutsche Gesellschaft für Internationale Zusammenarbeit (GIZ) GmbH et al., (eds.),
Sustainable Aging, https://doi.org/10.1007/978-3-662-69139-7_26

26.1   Fields of Action in Long-Term Care Insurance, Care Education, and Age-Friendly Cities and Communities

The establishment of a long-term care insurance system is one solution for countries with ageing societies to secure the long-term care of the elderly. Governments in respective countries, such as China and Germany, have taken different paths to implement these systems and this provides a chance for mutual exchange to benchmark and further improve the systems. Taking the legal framework of long-term care insurance in Germany as an example, Chinese experts in this book are calling for an integration of long-term care insurance into China's social security system. They argue that an integration would provide a unified path for the development of long-term care insurance and accelerate the implementation process. Besides the legal framework, financing mechanisms for the insurance system and the standardisation of an evaluation system greatly influence the system's success and could be an area of focus of further exchange. In the case of financing mechanisms, funding sources need to be sustainable considering the rising numbers of elderly. The financial burden can become too high, if payments for elderly care come out of the medical insurance which is currently the case in China. Furthermore, medical evaluation standards of the elderly need to be flexible enough to respond to diverse circumstances, such as care provided by relatives or care of younger people that developed disability due to accidents, as well as to future trends, such as dementia. Other pressing topics that need to be considered when developing insurance systems are the increasing number of empty nesters due to changing family structures as well as rural–urban disparities in infrastructure, medical services, or availability of professionals. In the case of China and Germany, the exchange on the establishment and adaptation of the insurance systems is still in its infancy and could be a future topic for Germany's international cooperation.

In the case of *elderly care education*, the situation is slightly different. The vocational education and training of nursing professionals has been an area of international cooperation in China and Southeast Asia as showcased in this book. For Germany and China, building up a nursing workforce is of greatest relevance, particularly since the pandemic highlighted the shortage of elderly care staff. Both countries are striving to improve the reputation and increase salaries in this sector to make the profession more appealing and bolster workforce supply. In the beginning, Germany tried to stem the shortfall by training elderly care assistants through certified short-term programmes which were eventually integrated into one of the officially recognised training programmes of the dual vocational education system. Faced with continuing high demands, in recent years the German government started recruiting trained nurses from countries with a surplus of professionals as shown in two project cases in this book. The Chinese government is taking a slightly different approach. To fill the gap, it has created new professions and is providing short-term training programmes, with a view to developing more long-term vocational training. Furthermore, the Chinese government has published guidelines for nurses to guarantee the wellbeing and safety of the people in need of care.

This book highlights the fact that traditional planning and design approaches applied in recent decades in China and Germany to adapt existing cities and communities to the changing needs of an ageing population are not sufficient to provide inclusive, healthy and pleasant living environments for today's elderly, i.e. *age-friendly cities and communities*. One reason is that elderly people live in and rely on diversified structures, with their own demands and preferences. Technical standards are important and form the basis for this more inclusive, integrated planning of the social, economic and environmental urban living, however as well as providing physically accessible urban spaces which meet the necessary technical standards, cities must also consider how to provide integrated living environments and spaces for recreation and leisure that enable the elderly to participate in society and live independently.

This demand is voiced by different contributors throughout the book and is in line with the ideas of the Sustainable Development Goals (SDGs) targeting inclusive living environments for all, regardless of their age, gender, religion, etc. To ensure inclusive planning, comprehensive concepts need to be put in place, including various actors from different levels (e.g. national and municipal), different disciplines (urban planning, architecture, social services, etc.) and different roles within the city (political and planning authorities, citizens and other stakeholders).

The approaches and concepts from Germany and China presented in this book form a good basis for future cooperation between both countries. The experience exchange on insurance systems, on training curricula, or on technical standards and participatory and community approaches to planning of urban spaces can be only beneficial in this exchange. The ideas provided can help address future challenges and transform them into opportunities to revise and improve existing systems as yet unprepared for the increasing global ageing trend. However, due to the COVID-19 pandemic, this exchange has practically come to a standstill. New suitable formats for international exchange need to be identified as an alternative to international experts meeting face to face. Nevertheless, whilst the pandemic has monopolised attention, which could potentially endanger progress already made, the crisis also offers chances for action: collapsing businesses force people to regroup and search for new opportunities, for example, within the nursing industry.

26.2 Lessons Learnt for International Cooperation

On a country level, addressing the comprehensive process of ageing societies requires intense cooperation and dialogue between the different national ministries of a country, but also between local authorities and society. On an international level, countries can benefit from an exchange on experiences and develop joint solutions. International cooperation institutions can act as facilitators of this cross-border exchange, not only by bringing together actors on different levels, but also by contributing knowledge gained through many decades of experience in the fields of social and human development to address demographics, health, social protection, family, education and youth, as well as ageing. Based on the articles and project cases in the present book, lessons learnt can be drawn and recommendations for action can be derived, as outlined below.

26.2.1 Foster Multi-level, Solution-Oriented Exchanges and Dialogues

Multi-level, solution-oriented exchanges and dialogues enable valuable insights into approaches applied in countries to mitigate these pressing issues, such as an ageing society, and they also offer the chance to identify issues for deeper collaboration. Such an exchange has been initiated between China and Germany at minister level in the health-sector, with the challenges of an ageing society being at the centre of the exchange, to discuss the transfer and scaling up of solutions applied in each individual country. A first conference, with participants from China, Japan and Germany, was held in Beijing in April 2019; it gathered experts from central and provincial government, associations, think tanks as well as the private sector. To find sustainable people-oriented solutions, however, the involvement of citizens is of crucial importance and should be given more consideration in future planning and discussion, as well as in the project implementation phase of Germany's international cooperation. The deeper involvement of different actors on different levels can further lead to increased ownership and sustainability of activities.

26.2.2 Foster Interdisciplinarity

Ensuring an inclusive life for all in which the elderly can live independently for as long as possible is a complex endeavour that covers various aspects of life, such as social and leisure activities, medical care, or employment. This endeavour thus demands an interdisciplinary approach by involving different disciplines and sectors. The success of international cooperation projects in creating a sustainable impact also relies on the application of a holistic, interdisciplinary approach. The present book brings together policy makers, practitioners and researchers from health management, education and urban planning as well as from the private sector. The World Health Organisation (WHO) provides guidance on which topics to consider when designing and implementing future cooperation projects, so for example to build age-friendly cities and communities, they suggest nine topics to be considered that touch upon various disciplines: urban and transport planning, architecture, social sciences, economics, health, education, engineering, and media sciences (Yeh et al., 2016).

26.2.3 Stronger Linking of Policy Making with Project Results

Some of the practical results showcased in this book are drawn from cooperation projects implemented by GIZ or from pilot projects implemented by the Chinese government. These results will serve the country's future policy making. This approach of pilot programmes shaping policy has been undertaken by the Chinese government for many years already. In the case of international cooperation projects, linking the project implementation more strongly to this process brings in international knowledge and experience. Policy recommendations drawn from practical experiences of implemented projects and pilots can foster bottom-up policy making, as these projects create stronger links to a diverse landscape of actors on the ground. Strong cooperation and deep exchange between cooperation projects, government think tanks and private sector form the basis for a continuous inclusion of results into policy making.

26.2.4 Ensure Upscaling of Results from China to Other Countries

China is a forerunner in many aspects related to the ageing industry, compared to many other developing countries and emerging markets, particularly in terms of technological and digital solutions. Other partner countries of German international cooperation will be confronted with these challenges several years later. Innovative results of the joint projects implemented by GIZ in China can therefore be used and transferred to other countries, for instance, in the framework of the Sino-German Centre for Sustainable Development (CSD), a joint initiative between the German Federal Ministry for Economic Cooperation and Development (BMZ) and the Ministry of Commerce of the People's Republic of China (MofCom) that supports developing countries to build an inclusive and sustainable future together.

26.2.5 Involvement of the Private Sector

The involvement of the private sector in establishing the elderly care sector is vital to cope with this pressing issue. Previously, providing for and taking care of the elderly was a task of the government and the family. With the number of elderly rising and the transformation of family structures, these responsibilities are changing. In the case of the care industry, for instance, the government is required to establish more comprehensive legal frameworks to meet increasing demand and relies on the private sector to provide technologies and services that can no longer be performed by the family. The engagement of the private sector can involve the establishment of elderly care institutions, the provision of training, the creation of insurance packages, the development of medical products, or tourism. However, by opening the market to the private sector, the government must retain the responsibility for ensuring the quality and safety of technologies and services by regulating and guiding the sector, so that, for example, care institutions are operated according to certain quality standards.

26.2.6 Digitalisation as a Future Topic

During the pandemic, the importance of digitalisation became even more apparent than before, especially during the lockdowns imposed in many countries. Digitalisation helped keep societies going and enabled the partial continuation of crucial activities, such as education, the supply of daily necessities, and work. Digitalisation can help societies prepare for future pandemics and mitigate the ageing challenge; technical devices can easily document health conditions to improve diagnosis, and better align care services, or support nurses when monitoring patients' health conditions. Installed sensors can report changes of health conditions or patients' critical statuses to secure immediate response. In the field of care education, applications equipped with virtual reality can also be included in practical training programmes for nurses. Digitalisation thus needs a stronger emphasis in the future, in governmental policies, in the industry as well as in international cooperation. China's different industries, for instance, are eagerly developing digital solutions on a very broad scale, which will likely also prove valuable for other countries when transforming their professional sectors related to the elderly and their lives.

References

DW, Deutsche Welle. (2020). *Germany's birth rate drops, confirming dramatic predictions for the whole world*. Retrieved July 31, 2020, from https://www.dw.com/en/demography-german-bir thrate-down-in-coronavirus-pandemic/a-54395345. Accessed 02 April 2021

Statista. (2021). *Ageing population in China: Statistics and facts*. Retrieved April 06, 2021, from https://www.statista.com/topics/6000/ageing-population-in-china/. Accessed 07 April 2021

Yeh, C. C., Walsh, J., Spensley, C., & Wallhagen, M. (2016). Building inclusion: Toward an ageing- and disability-friendly city. *American Journal of Public Health, 106*(11), 1947–1949. https://doi.org/10.2105/AJPH.2016.303435

Marie Peters is an urbanisation expert with more than 10 years of work experience in East Asia, Southeast Asia and Europe. She trained as an urban geographer at the University of Cologne, Germany, with a focus on sustainable urbanisation, climate-risk resilience, and urban planning and development. She holds a Ph.D. in Geography from the University of Cologne. Her Ph.D. research covered Chinese inter-city competition for talents in the mega-urban region of the Pearl River Delta, China. She has also published on international migration and climate adaptation as well as on the dynamics of industrial upgrading. Her general academic interests include the role of migration within urban transformation and climate adaptation. Marie joined Deutsche Gesellschaft für Internationale Zusammenarbeit (GIZ) GmbH in 2015, where she has been working as a policy advisor for different projects related to urbanisation and transport in Germany and China. She is currently based in Berlin serving as advisor to the German Council for Sustainable Development.

Sabine Porsche graduated in Cultural Anthropology from Philipps-University in Marburg. She joined Deutsche Gesellschaft für Internationale Zusammenarbeit (GIZ) in Beijing in 2017, where she headed the team Society and Labour. Topics of interest are elderly care and education. From 2007 to 2016 Sabine held different positions at Tongji University in Shanghai. She held the position as German Vice-Director of the Sino-German University of Applied Sciences and besides established the DAAD Career Academy.